Minimizing Spurious Tones in Digital Delta-Sigma Modulators

ANALOG CIRCUITS AND SIGNAL PROCESSING

Series Editors: Mohammed Ismail
Mohamad Sawan

For other titles published in this series, go to
http://www.springer.com/series/7381

Kaveh Hosseini · Michael Peter Kennedy

Minimizing Spurious Tones in Digital Delta-Sigma Modulators

 Springer

Kaveh Hosseini
Cypress Semiconductor
Cork, Ireland
kawe.hosseini@gmail.com

Michael Peter Kennedy
University College Cork
Cork, Ireland
peter.kennedy@ucc.ie

ISBN 978-1-4614-2921-0   ISBN 978-1-4614-0094-3 (eBook)
DOI 10.1007/978-1-4614-0094-3
Springer New York Dordrecht Heidelberg London

Printed on acid-free paper

Springer is part of Springer Science+Business Media (www.springer.com)

To Monireh, Mohammad Saleh, Samira (K.H.)
To Rossana (M.P.K.)

Preface

Analog Delta Sigma Modulators (ADSM) have been extensively analyzed and used in the context of analog-to-digital conversion; however, less attention has been paid to Digital Delta Sigma Modulators (DDSM) which are commonly used in digital-to-analog conversion and fractional-N frequency synthesis. Motivated by this fact, combined with their widespread use in wireless transceivers, we aim to demystify an important aspect of some popular DDSM structures, namely the existence of spurious tones due to the inherent periodicity of signals in DDSMs with constant inputs. The architectures under investigation include Multi-stAge noise SHaping (MASH), Single Quantizer (SQ) and Error Feedback (EF) DDSMs.

A DDSM is a finite state machine (FSM); it is implemented using finite precision arithmetic units and the number of available states is finite. A deterministic FSM has a deterministic rule for transitioning from each state to the next. If the input is constant, the most complex behavior the DDSM can exhibit is a trajectory that visits each state once before repeating; in fact, the output must always be constant or periodic. Therefore, the DDSM always produces a periodic output signal (a cycle) when the input is constant. Furthermore, the quantization error signal (commonly called the quantization noise) is also periodic in this case.

When the length of the cycle is short, the average power of the quantization noise in the DDSM is spread over a small number of discrete tones. According to Parseval's relation, the shorter the cycle length is, the fewer tones there are, and consequently the higher the power per tone. Undesirable tones of this type in DDSMs are called spurious tones or spurs. If the cycle length is sufficiently large and the quantization noise samples between cycles are sufficiently randomized, the shaped output quantization noise spectrum becomes indistinguishable in practice from a continuous spectrum.

There are two classes of techniques for maximizing cycle lengths in DDSMs: "stochastic" and deterministic. The "stochastic" approach to maximizing cycle lengths is to use a pseudo-random dither sequence to disrupt periodic cycles. Dithering breaks up the cycles and increases the effective cycle length, resulting in smooth noise-shaped spectra. While the "stochastic" solution increases the cycle length, as required, it inherently adds noise to the spectrum; care must be taken to minimize the effects of this additional noise. By contrast, the objective of deterministic approaches is to guarantee maximum cycle lengths by design, without the need for an external dithering signal. In this book, we focus primarily on deterministic techniques.

In Chapter 1, we explain briefly the concept of noise shaping in delta-sigma (DS) modulators. Then, we explain the problem which we address. The main contributions of the book are summarized at the end of this chapter.

In Chapter 2, we describe delta-sigma modulation (DSM) and provide an overview of two applications for DDSMs, namely digital-to-analog converters and fractional-N frequency synthesizers.

In Chapter 3, we review popular dithering techniques. Then, considering deterministic techniques, we provide an overview of some deterministic techniques for maximizing cycle lengths and we provide mathematical proofs concerning these.

In Chapters 4 and 5, we describe a deterministic technique for maximizing cycle lengths in MASH, SQ and EFM DDSMs.

This work was supported in part by Science Foundation Ireland under grants 02/IN.1/I45 and 08/IN1/I854.

Cork, Ireland   Kaveh Hosseini
March 2011   M. Peter Kennedy

Contents

Acronyms

AC	Alternating Current; in this book, it denotes a time-varying signal
ADC	Analog-to-Digital Converter
ADSL	Asymmetric Digital Subscriber Line
ADSM	Analog Delta Sigma Modulator
CP	Charge Pump
CMOS	Complementary Metal Oxide Semiconductor
CT	Continuous Time
CS	Current Steering
DAC	Digital-to-Analog Converter
DC	Direct Current; here it denotes a constant signal
DCS	Digital Cellular System
DDS	Direct Digital Synthesis
DDSM	Digital Delta Sigma Modulator
DFT	Discrete Fourier Transform
DTFS	Discrete Time Fourier Series
DR	Dynamic Range
DT	Discrete Time
DS	Delta Sigma
DSM	Delta Sigma Modulation
DTFS	Discrete Time Fourier Series
EFM	Error Feedback Modulator
FD	Frequency Divider
FFT	Fast Fourier Transform
FPGA	Field-Programmable Gate Array
FS	Frequency Synthesizer
FSM	Finite State Machine
GCD	Greatest Common Divisor
IF	Interpolation Filter
LC	Limit Cycle
LFSR	Linear Feedback Shift Register
LO	Local Oscillator
LPF	Low Pass Filter

LSB	Least Significant Bit
MASH	Multi-stAge noise SHaping
MSB	Most Significant Bit
NL	Noise shaping Loop
NTF	Noise Transfer Function
OSR	OverSampling Ratio
PD	Phase Detector
PDF	Probability Density Function
PFD	Phase Frequency Detector
PLL	Phase Locked Loop
PRBS	Pseudo Random Binary Sequence
PSD	Power Spectral Density
QNOB	Quantizer Number Of Bits
RBW	Resolution BandWidth
RC	Resistor Capacitor
RF	Radio Frequency
SFDR	Spurious Free Dynamic Range
SNR	Signal-to-Noise Ratio
SQNR	Signal-to-Quantization-Noise Ratio
SQ	Single Quantizer
SC	Switched Capacitor
STF	Signal Transfer Function
UMTS	Universal Mobile Telecommunications System
VCO	Voltage Controlled Oscillator
VDSL	Very high bitrate Digital Subscriber Line
WLAN	Wireless Local Area Network

Chapter 1
Introduction

Delta-sigma modulation is a technique for improving the effective resolution of a quantizer by oversampling and noise shaping. The idea is to suppress the quantization noise of a quantizer around the desired signal band by using negative feedback. The suppression of the quantization noise around the signal band is called noise shaping. The earliest descriptions of the technique were reported by Jager [1] and Cutler [2]. Inose et al. [3] reported a delta-sigma analog-to-digital converter (ADC) system (see Fig. 1.1) that uses a discrete-time filter ($F(z)$) and a 1-bit quantizer (denoted ($Q(\cdot)$)) in the forward path, and a 1-bit digital-to-analog converter (DAC) in the feedback loop of a negative feedback system. They called the system a delta-sigma modulator, where delta denotes the front end subtraction block of the negative feedback system, and sigma refers to the summation performed in the feedforward filter. The output y of the modulator contains information of the original input signal x plus the filtered quantization noise introduced by the quantizer. The noise is filtered in such a way that its power is attenuated in the signal band.

In order to illustrate the concept of noise shaping, we refer to Fig. 1.2 that represents the operation of a delta-sigma modulator schematically in the frequency domain. Assume that we have applied to the input of the modulator an analog or a high resolution digital signal[1] sampled with the rate of f_s, as shown in Fig. 1.2a. The modulator delivers an output signal with an amplitude resolution that is much

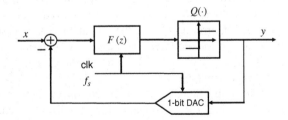

Fig. 1.1 Block diagram of a 1-bit delta-sigma modulator

---

[1] If the input is digital, the DAC in the feedback path is not required and the filter F is implemented digitally.

K. Hosseini, M.P. Kennedy, *Minimizing Spurious Tones in Digital Delta-Sigma Modulators*, Analog Circuits and Signal Processing, DOI 10.1007/978-1-4614-0094-3_1, © Springer Science+Business Media, LLC 2011

Fig. 1.2 Illustration of quantization noise shaping in a low-pass discrete-time delta-sigma modulator clocked at f_s. (**a**) The input spectrum. (**b**) The output spectrum of the modulator; it contains the input spectrum in addition to the filtered quantization noise. (**c**) An ideal continuous-time low pass filter is applied to the modulator output. (**d**) The filtered output spectrum; the contribution of the quantization noise component in the signal band is small

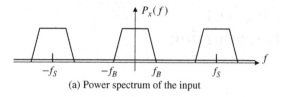

(a) Power spectrum of the input

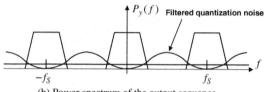

(b) Power spectrum of the output sequence

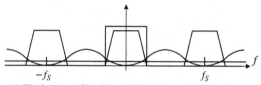

(c) The low pass filter is applied to the modulator output

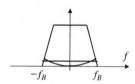

(d) The low pass filter output spectrum (enlarged)

lower than the input resolution.[2] As shown in Fig. 1.2b, the spectrum of the output contains the original spectrum of the input signal plus the filtered quantization noise.

In producing a coarse output signal from a high resolution input signal, a quantization error is generated. Ideally, the quantization error is whitened and filtered by the modulator so that its power is moved away from the signal. The shaped quantization error appears in the output spectrum, as shown in Fig. 1.2b. Note that the power of the quantization noise is concentrated away from the signal band f_B and it is concentrated around $\frac{f_s}{2}$, thereby maximizing the signal to quantization noise ratio in the signal band.

If the output signal of the modulator is applied to an ideal continuous-time low-pass filter (Fig. 1.2c) that passes only the low frequency content, then the original signal can be recovered from the low resolution output signal with a high signal to quantization noise ratio. Figure 1.2d shows the spectrum at the output of the

---

[2] The output signal is typically quantized to one to three bits.

low-pass filter (enlarged). If the noise power of the shaped quantization noise in the signal band is negligible compared to that of the original quantized signal x, then the filtered output signal y can have a signal to noise ratio that is close to that of the original quantized input x. The high resolution signal x can thus be transformed into a low resolution signal y with negligible degradation in SNR.

Since the invention of the basic delta-sigma architecture, DS modulation has been used extensively and many architectures have emerged, with one-bit and multi-bit quantizers, single and multiple loops, various orders of filter, continuous-time and discrete-time, analog and digital implementations, depending on the application and the required performance [4].

Understanding and analyzing issues such as stability and spurious tone generation in delta-sigma modulators is difficult [4]. This is because the modulators are negative feedback systems which include strongly nonlinear quantizers; therefore, results from linear control theory do not apply to these systems. Despite the fact that linear analysis is often used to explain the overall behavior of a delta-sigma modulator and that it can give an estimate of the signal to quantization noise ratio, it is not the best tool for analyzing problems resulting from the nonlinear nature of the system. In that case, designers must rely on extensive computer simulations and, where available, rigorous nonlinear analyses [4–6]. In addition, the analysis of delta-sigma modulators becomes tedious when higher order filters (order greater than one) are used in the implementation [4–6]. A general study that covers all issues and all types of implementation is difficult; authors often choose a specific topology with a specific type of implementation in order to simplify the analysis.

Analog DSMs have been analyzed and used extensively in the context of ADCs; however, less attention has been paid to digital delta-sigma modulators (DDSM). DDSMs are widely used in fractional-N frequency synthesizers and DACs. In this work, we consider the DDSM. In a DDSM, the input is digital and the filters are implemented digitally.

The performance of a delta-sigma modulator is usually evaluated by observing its output power spectral density (PSD). Ideally, the PSD is free of undesirable spurious tones (spurs); in practice, most delta-sigma modulators have PSDs which contain spurs like those illustrated in Fig. 1.3. The existence of spurs in the output PSD can seriously degrade the spurious free dynamic range (SFDR) of a modulator.

We aim to demystify an important aspect of some particular DDSM structures, namely the existence of spurs resulting from the inherent periodicity of DDSMs with constant inputs. The architectures we consider include Multi staAge noise SHaping (MASH), single quantizer (SQ) and error feedback modulator (EFM) topologies, all

Fig. 1.3 Spurious tones appear in the output PSD of a delta-sigma modulator

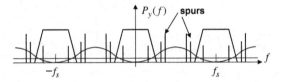

of which have an all-pass signal transfer function (STF) and a noise transfer function (NTF) of the form $(1 - z^{-1})^l$.

Firstly, we need to clarify the term "inherent periodicity". A DDSM is a finite state machine (FSM) because it is implemented using finite precision arithmetic units and the number of available states is finite. This deterministic FSM has a unique deterministic rule for transitioning from each state to the next. If the input is constant, the most complex behavior the DDSM can exhibit is a trajectory that visits each state once before repeating; in fact, the output must always be constant or periodic. Therefore, the DDSM always produces a periodic output signal (called a cycle) when the input is constant. In general, the length of the cycle may depend on the input, the initial conditions, and the architecture of the DDSM.

When the cycle length is short, the output power is spread over a small number of discrete tones. According to Parseval's relation [7], the shorter the cycle length is, the fewer tones there are, and consequently the higher the power per tone. The cycle length depends on the input, the initial conditions, and the architecture of the DDSM. In some cases, the cycle length may become very small, resulting in large spurs in the output power spectrum.

Here we provide an example to illustrate how the dependence of the period on the initial conditions is reflected in the spectrum. A third order MASH 1-1-1 DDSM shown in Fig. 1.4 with the accumulator wordlength $n_0 = 14$ is simulated under two different conditions. The input to the modulator in both cases is constant, namely $X = 256$. Figure 1.5a shows the PSD of the output of the DDSM when the initial conditions of the three stages are: $s_1[0] = s_2[0] = s_3[0] = 0$. Figure 1.5b shows the PSD when $s_1[0] = 1$ and $s_2[0] = s_3[0] = 0$. The dramatic difference between the two cases results solely from the difference in the initial condition of the first stage. As we will see in Chapter 3, the cycle lengths associated with Fig. 1.5a, b are 2^7 and 2^{15}, respectively. Note that the spectrum associated with the odd initial condition can be interpolated by a smooth curve pushing the quantization noise power gently toward the high frequencies. By contrast, the spectrum associated with the zero initial condition contains some high power tones resulting from the relatively short cycle length of the output sequence. This is an example of a nonlinear dynamical system which exhibits multiple (coexisting) steady state solutions. Starting from one initial condition yields a periodic solution with a long period while another initial state leads to a solution with a much smaller period and consequently higher power tones in the output spectrum.

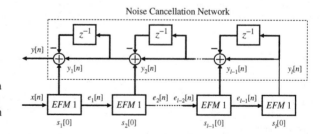

Fig. 1.4 Block diagram of a MASH DDSM. The system will be described in detail in the next chapter

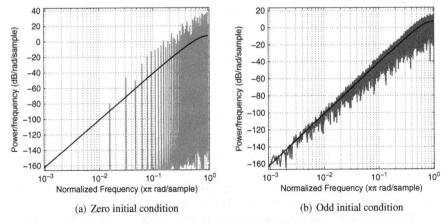

(a) Zero initial condition   (b) Odd initial condition

Fig. 1.5 The effect of short cycles in the 14-bit MASH 1-1-1 DDSM shown in Fig. 1.4: (**a**) zero initial condition $s_1[0] = s_2[0] = s_3[0] = 0$; (**b**) odd initial condition, $s_1[0] = 1$ and $s_2[0] = s_3[0] = 0$. The constant input is $X = 256$ in both cases. The *solid curve* shows the idealized shaped white quantization noise

There are two classes of techniques for maximizing cycle lengths in DDSMs: "stochastic" and deterministic. The "stochastic" approach to maximizing cycle lengths is to use a "random"[3] dither sequence to disrupt periodic cycles [4, 8–10]. Dithering breaks up the cycles and increases the effective cycle length, resulting in smooth noise-shaped spectra. While the "stochastic" solution increases the cycle length, as required, it inherently adds noise to the spectrum; care must be taken to minimize the contribution of this additional noise. By contrast, the deterministic approach to whitening the quantization noise is to guarantee maximum cycle lengths by design, without the need for an external dithering signal. In this work, we focus primarily on deterministic techniques.

In the next chapter, we present an overview of DS modulators and illustrate some common architectures. Then, we describe two applications for DDSMs, namely digital-to-analog converters and fractional-N frequency synthesizers.

1.1 Contributions of This Book

In this book, we consider several DDSM architectures, including MASH, EFM and SQ-DDSM modulators, with a view to predicting and maximizing their cycle lengths. The main contributions of this work are as follows:

1. Borkowski et al. have found specific sets of initial conditions that, when applied to MASH and EFM DDSMs, yield cycles of specific lengths [11, 12]. They

---

[3] In practice, the "random" sequence might be generated by a pseudo-random or chaotic source.

suggest that the cycle lengths can be guaranteed to be larger than some minimum values provided that certain conditions are met. In particular, the lengths of the cycles can be increased if initial conditions and/or inputs are constrained in special ways and the modulator's wordlengths are increased. The results in [11] are empirical in nature. In Chapter 3, we provide a proof of these results in the special cases of MASH modulators with orders 1, 2 and 3 [13].

2. Following our proof of the results given in [11], we provide exact mathematical expressions [14] for MASH modulators of order 1, 2 and 3 with constant inputs that enable one to determine the cycle lengths as a function of the word length, the initial conditions, and the input. This allows a designer to predict the locations of spurious tones analytically.

3. Level and Camino invented a MASH DDSM which uses a prime number for the moduli of the internal quantizers [15]. In Chapter 3 we prove that using a prime number as the modulus in the internal quantizers guarantees a cycle length that is equal to this prime number [16]. In addition, we prove that the cycle length is independent of the initial conditions and the input when the latter is constant.

4. Our novel MASH modulator, denoted HK-MASH, will be described in Chapter 4. It yields maximized cycle lengths that are independent of the initial conditions and the input values. In contrast to the modulator described in [15], the HK-MASH modulator does not use a prime number to implement the modulus of the quantizer. Rather, it uses conventional power-of-two quantizers with a modulus $M = 2^{n_0}$ where n_0 is an integer; this is simpler to implement in practice. Instead of implementing a more complex prime modulus quantizer, we include a delayed feedback path with a specially chosen coefficient that adds a small number a to the state of the modulator each time the quantizer crosses its threshold value. The difference between the step-size and the coefficient in the feedback path is a prime number. We prove mathematically that the cycle length can be increased by increasing the number of stages in the MASH modulator. If the cycle length is $(M - a)$ for one stage, the cycle length in the HK-MASH DDSM becomes $(M - a)^l$, where l is the order of the MASH modulator. We showed in [16] that the cycle length in [15] remains the same, namely $M_p = M - a$ for any order of MASH modulator of that type, while our structure achieves $(M - a)^l$.

5. In Chapter 5, we extend the idea that underpins the HK-MASH modulator and show that the same idea can be applied to other classes of modulators including higher order multi-bit EFM and SQ-DDSM modulators.

Simulink and MATLAB models and code are presented in Chapters 2, 3, 4, and 5 to enable the reader to reproduce the results in this work and to explore further. We hope that these examples will also be helpful for first-time designers of DDSMs. All MATLAB and Simulink files described in this book are available for download from: http://cas.tyndall.ie.

Chapter 2
DDSM and Applications

2.1 Principles of Delta-Sigma Modulation

In order to explain the concept of noise shaping in detail, we start with a stand-alone quantizer (see Fig. 2.1a) with a small number of bits that maps the amplitude of an analog input signal to only a few output levels. We show examples of two types of quantizers in Fig. 2.2, namely mid-tread and mid-rise quantizers [6]. These map an analog signal v to a low resolution output signal y having only five or four discrete levels, respectively. Δ denotes the quantizer step size in Fig. 2.1a and k is a gain factor, which is 1 in the two examples in Fig. 2.2.

Consider the quantizers shown in Fig. 2.2, which have a step-size Δ. If the input to the mid-tread quantizer is greater than $5\frac{\Delta}{2}$ or less than $-5\frac{\Delta}{2}$, the quantizer saturates at its maximum and minimum output values of 2Δ and -2Δ, respectively. If the input lies between these values (i.e., $|v| \leq 5\frac{\Delta}{2}$), the quantizer is said to operate in the no-overload range. Similarly, the mid-rise quantizer in this example is in its no-overload range when $|v| \leq 2\Delta$.

This coarse input–output mapping (quantization) introduces a quantization error (see Fig. 2.1b) defined by

$$e_q = y - kv, \tag{2.1}$$

where k is a gain factor. In the literature, the quantization error is conventionally called quantization noise [4, 17].

For the quantizers shown in Fig. 2.2, we have illustrated the quantization error e_q as a function of the input v (see Fig. 2.2b, d). In the no-overload range, the quantization noise is bounded in the range $\left[-\frac{\Delta}{2}, \frac{\Delta}{2}\right]$.

Fig. 2.1 (a) A quantizer block diagram and (b) its linearized model

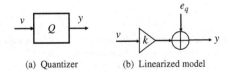

(a) Quantizer　　　(b) Linearized model

K. Hosseini, M.P. Kennedy, *Minimizing Spurious Tones in Digital Delta-Sigma Modulators*, Analog Circuits and Signal Processing, DOI 10.1007/978-1-4614-0094-3_2, © Springer Science+Business Media, LLC 2011

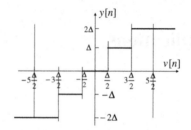

(a) A mid-tread quantizer

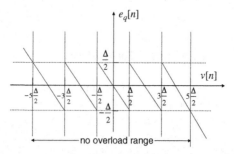

(b) Quantization error e_q

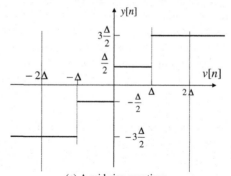

(c) A mid-rise quantizer

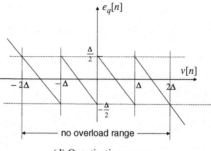

(d) Quantization error e_q

Fig. 2.2 (**a**) Transfer characteristic of a coarse mid-tread quantizer with five output levels. (**b**) The quantization error e_q of the quantizer defined by $e_q = y - kv$ with $k = 1$. (**c**) Transfer characteristic of a coarse mid-rise quantizer with four output levels. (**d**) The quantization error e_q of the quantizer defined by $e_q = y - kv$ with $k = 1$. Δ is the step-size of the quantizer

The signal-to-quantization-noise ratio (SQNR) is defined as the ratio of the signal power to the power of the quantization noise in a given frequency range, which is typically the signal bandwidth. For a given signal power, coarse quantization (using only a few levels) results in a small SQNR. To increase the SQNR for a given signal power, one must decrease the quantization noise power. This can be achieved by increasing the number of levels in the quantizer. Thus, the larger the required SQNR, the greater the required resolution of the quantizer.

Alternatively, we can use the concept of noise shaping and place the coarse quantizer in a delta-sigma modulator loop, as shown in Fig. 2.3a. Qualitatively, the delta-sigma modulator attenuates the quantization noise power in the signal band and amplifies the out-of-band power. If appropriate filtering is subsequently applied, the out-of-band quantization noise can be attenuated significantly. As we will see, the attenuation of the quantization noise in the signal band results in a larger SQNR compared with the SQNR of a stand-alone coarse quantizer.

We assume the sampling frequency f_s of the modulator is much greater than twice the bandwidth of the input signal. If a discrete-time signal is sampled at a rate that is much greater than twice its bandwidth, it is said to be oversampled. The oversampling ratio OSR is defined as:

$$\text{OSR} = \frac{f_s}{2f_B}, \tag{2.2}$$

where f_s is the sampling frequency and f_B is the largest frequency component in the signal spectrum. $2f_B$ corresponds to the Nyquist rate [7]; the OSR is the factor by which the sampling frequency exceeds to the Nyquist rate. For example, if $f_B = 44\,\text{kHz}$ and $f_s = 5.632\,\text{MHz}$, then the Nyquist rate is $88\,\text{kHz}$ and OSR$= 64$.

In data converter applications, oversampling and the noise shaping property of the DSM can be used together to achieve a high SQNR using a relatively coarse quantizer; these systems are referred to as oversampled delta-sigma converters [4].

As mentioned earlier, the delta-sigma modulator places the quantizer in a negative feedback loop, as shown in Fig. 2.3a. This architecture contains discrete-time feedback and feedforward filters $G(z)$ and $F(z)$, respectively. In the following, we explain the roles of these filters.

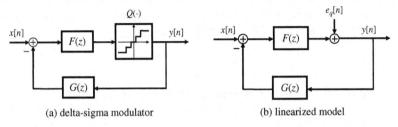

(a) delta-sigma modulator   (b) linearized model

Fig. 2.3 Block diagrams of (**a**) a single-quantizer discrete-time delta-sigma modulator and (**b**) its linearized model with the quantizer gain factor $k = 1$. The input has been oversampled with an oversampling ratio OSR

We can build a simplified linear model of the modulator by replacing the quantizer with a gain factor[1] k and an additive signal source e_q, as shown in Fig. 2.3b [4]. Taking the z transforms of the signals in Fig. 2.3b, $Y(z)$ can be written in terms of the z-transforms of the input signal $X(z)$ and the quantization noise $E_q(z)$ as follows:

$$Y(z) = \text{STF}(z)X(z) + \text{NTF}(z)E_q(z), \qquad (2.3)$$

where

$$\text{STF}(z) = \frac{F(z)}{1 + F(z)G(z)}, \qquad (2.4)$$

$$\text{NTF}(z) = \frac{1}{1 + F(z)G(z)}. \qquad (2.5)$$

STF(z) and NTF(z) are the signal and noise transfer functions, respectively, of the linearized system. These transfer functions are usually designed such that the input signal is not attenuated by the system while the quantization noise is strongly attenuated in the signal band.

In this book, we deal with low pass signals; therefore, the required NTF is high pass in nature, meaning that it attenuates the quantization noise at low frequencies and passes the quantization noise at high frequencies. For example, assume that

$$F(z) = \frac{1}{1 - z^{-1}}, \qquad (2.6)$$

$$G(z) = z^{-1}, \qquad (2.7)$$

where $F(z)$ is the transfer function of an integrator (or an accumulator in a digital implementation) and $G(z)$ is a simple delay. With these filters, we obtain

$$\text{STF}(z) = 1, \qquad (2.8)$$

$$\text{NTF}(z) = 1 - z^{-1}. \qquad (2.9)$$

The NTF is a first order high pass filter, as we will see below. In order to understand the behavior of the system in the frequency domain, we write $z = e^{j\omega}$ and calculate the squared magnitudes of the NTF and STF. Note that

$$|\text{STF}(z)|^2_{z=e^{j\omega}} = 1,$$

so the signal passes without any filtering from the input to the output.

---

[1] In this case, $k = 1$.

For the NTF, we can write

$$\mathrm{NTF}(e^{j\omega}) = 1 - e^{-j\omega}$$
$$= 1 - \cos(\omega) + j\sin(\omega),$$

and

$$\left|\mathrm{NTF}(e^{j\omega})\right|^2 = (1 - \cos(\omega))^2 + \sin^2(\omega)$$
$$= 2 - 2\cos(\omega)$$
$$= 2\left(2\sin^2\left(\frac{\omega}{2}\right)\right)$$
$$= \left(2\sin\left(\frac{\omega}{2}\right)\right)^2. \tag{2.10}$$

The magnitude response of the noise transfer function is thus $2\left(\sin\left(\frac{\omega}{2}\right)\right)$. The peak of this function occurs at $\omega = \pi$ with a magnitude of 2. We are interested in low pass signals so let us examine the function around $\omega \approx 0$. For small ω,

$$2\left(\sin\left(\frac{\omega}{2}\right)\right) \approx 2\left(\frac{\omega}{2}\right) = \omega.$$

Decreasing ω decreases the magnitude of the frequency response. At zero frequency, the magnitude of the frequency response becomes zero. We conclude that the delta-sigma modulator in this example implements quantization noise shaping with a high pass characteristic, meaning that it attenuates the noise at low frequencies ($\omega \approx 0$) and amplifies the noise at high frequencies[2] ($\omega \approx \pi$).

Figure 2.4a shows $\left|\mathrm{NTF}(e^{j\omega})\right|^2$ with ω in the range [0 2π] for two cases: (i) $\mathrm{NTF}(z) = (1 - z^{-1})$; (ii) $\mathrm{NTF}(z) = \left(1 - z^{-1}\right)^2$. Note that the sampling frequency f_s maps to 2π.

2.1.1 SQNR

Now that we have explained the noise shaping property of the modulator with the help of the linearized model in the first order modulator, we would like to calculate its SQNR by making simplifying assumptions about the statistics of the quantization noise. In practice, these assumptions may not hold in all cases. In particular, as we will see in the next chapters, one of the drawbacks of the white noise approximation is that it does not allow us to predict spurious tones in DSMs. Nevertheless, it allows one to estimate how the SQNR can be improved by increasing the modulator order

---

[2] Note that $\omega = 2\pi\frac{f}{f_s}$ [7]; therefore, the sampling frequency f_s is mapped to 2π.

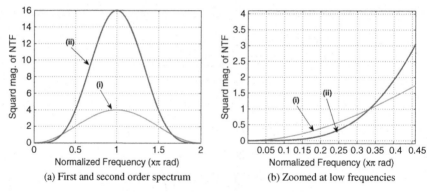

Fig. 2.4 (**a**) The squared magnitude of the noise transfer function in a first order [plot (*i*)] and a second order [plot (*ii*)] delta-sigma modulator. (**b**) Zooms of the plots shown in (**a**) at low frequencies. The second order modulator provides greater attenuation of the quantization noise at low frequencies, but amplifies it at high frequencies

and the OSR. When we refer to the white noise approximation throughout this book, we will assume that the following assumptions are valid unless otherwise stated.

We assume that:

- the quantization noise is uniformly distributed in the range $\left[-\frac{\Delta}{2} \; \frac{\Delta}{2}\right]$ and has zero mean;
- it is independent of the input;
- delayed versions of the noise are uncorrelated with each other.

The last condition implies that $R_{ee}[l_g] = \sigma_e^2 \delta[l_g]$ [7], where $R_{ee}[l_g]$ is the autocorrelation function of e_q, σ_e^2 is the variance of e_q, and l_g is a given lag.

First we calculate the power of the quantization noise. The variance of e_q (the average power of e_q) is calculated as follows [7]

$$\sigma_e^2 = E[e_q^2] = \int_{-\frac{\Delta}{2}}^{\frac{\Delta}{2}} f_{pdf}(e_q) e_q^2 de_q = \int_{-\frac{\Delta}{2}}^{\frac{\Delta}{2}} \frac{1}{\Delta} e_q^2 de_q = \frac{\Delta^2}{12},$$

where $f_{pdf}(e_q) = \frac{1}{\Delta}$ is the probability density function of e_q with the uniform distribution shown in Fig. 2.5a.

The autocorrelation function of e_q for a lag l_g is given by

$$R_{ee}[l_g] = \sigma_e^2 \delta[l_g],$$

which has a nonzero value of $\frac{\Delta^2}{12}$ at zero lag; otherwise, it is zero. The power spectral density of e_q is the discrete-time Fourier transform of the autocorrelation function [7]. This results in a white power spectral density $P_{eq}(e^{j\omega}) = \sigma_e^2$, as shown in Fig. 2.5b, with a constant amplitude of $\sigma_e^2 = \frac{\Delta^2}{12}$.

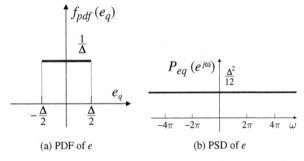

(a) PDF of e   (b) PSD of e

Fig. 2.5 (a) Probability density function (PDF) of the quantization noise uniformly distributed in $\left[-\frac{\Delta}{2}\ \frac{\Delta}{2}\right]$. (b) Power spectral density (PSD) of the quantization noise. The PSD is the discrete-time Fourier transform of the autocorrelation function $R_{ee}[l_g]$ [7]

If the above assumptions hold, the filtered quantization noise has the same shape as the NTF with a scaling factor of σ_e^2. Hence,

$$P_{e_{sh}}\left(e^{j\omega}\right) = |\text{NTF}|^2 \left|P_{eq}\left(e^{j\omega}\right)\right| = \left(2\sin\left(\frac{\omega}{2}\right)\right)^2 \sigma_e^2,$$

where $P_{e_{sh}}$ is the PSD of the shaped quantization noise. Next, we estimate the SQNR for this first order modulator when the white noise approximation is valid. Assuming that the sampling frequency is $f_s = \text{OSR} \times 2f_B$ and that f_s is mapped at 2π, the upper edge of the signal band will be located at $\frac{\pi}{\text{OSR}}$. The total quantization noise power within the signal bandwidth $\left[-\frac{\pi}{\text{OSR}}\ \frac{\pi}{\text{OSR}}\right]$ is calculated as follows:

$$\sigma_{e_{sh}}^2 = 2\frac{1}{2\pi}\int_0^{\frac{\pi}{\text{OSR}}} \sigma_e^2\left(2\sin\left(\frac{\omega}{2}\right)\right)^2 d\omega \tag{2.11}$$

$$= \frac{8}{\pi}\sigma_e^2\left(\frac{\pi}{4\text{OSR}} - \frac{1}{4}\sin\left(\frac{\pi}{\text{OSR}}\right)\right). \tag{2.12}$$

Expanding the second term as a Taylor series, we obtain

$$\sin\left(\frac{\pi}{\text{OSR}}\right) \approx \frac{\pi}{\text{OSR}} - \frac{\left(\frac{\pi}{\text{OSR}}\right)^3}{6}, \quad \text{for } \frac{\pi}{\text{OSR}} << 1. \tag{2.13}$$

Substituting (2.13) into (2.12) yields

$$\sigma_{e_{sh}}^2 \approx \sigma_e^2 \frac{\pi^2}{3}\frac{1}{(\text{OSR})^3}. \tag{2.14}$$

The SQNR is defined by

$$\text{SQNR} = \frac{\text{signal power}}{\text{quantization-noise power in the signal band}} \tag{2.15}$$

$$= \frac{P_{x,ave}}{\sigma_{e_{sh}}^2} \tag{2.16}$$

$$\approx \frac{P_{x,ave}}{\sigma_e^2} \frac{3}{\pi^2} \text{OSR}^3 \tag{2.17}$$

It is common to measure SQNR in decibels. Thus,

$$\text{SQNR}_{dB} \approx 10 \log_{10} \frac{P_{x,ave}}{\sigma_e^2} - 10 \log_{10}\left(\frac{\pi^2}{3}\right) + 30 \log_{10}(\text{OSR}). \tag{2.18}$$

If the oversampled signal is applied directly to the stand-alone quantizer without the delta-sigma loop, the SQNR in dB is given by [5]

$$\text{SQNR}_{dB} = 10 \log_{10} \frac{P_{x,ave}}{\sigma_e^2} + 10 \log_{10}(\text{OSR}). \tag{2.19}$$

Note that doubling the OSR increases the SQNR by 3 dB in the stand-alone quantizer, while doubling the OSR results in an increase of approximately 9 dB in the SQNR of a first order delta-sigma modulator. This is why noise shaping helps to improve the resolution of a quantizer.

We can use higher order modulators to obtain a larger SQNR. In the case of a second order modulator (see Fig. 2.4), the SQNR is given by [5]

$$\text{SQNR}_{dB} \approx 10 \log_{10} \frac{P_{x,ave}}{\sigma_e^2} - 10 \log_{10}\left(\frac{\pi^4}{5}\right) + 50 \log_{10}(\text{OSR}). \tag{2.20}$$

Every doubling of the OSR in the second order modulator results in an increase of approximately 15 dB in the SQNR. In general, for an lth order modulator with $\text{NTF}(z) = \left(1 - z^{-1}\right)^l$, the SQNR increases by $(6l + 3)$ dB for every doubling of the OSR [6].

$\text{NTF}(z) = \left(1 - z^{-1}\right)^l$ is not the only possible filter choice in the design of delta-sigma modulators. Locating the zeros and poles of the NTF by design allows one to prescribe the inband SQNR, the modulator stability, and the spectral performance of the modulator (see, for example [6, Chap. 4]).

In this section, we reviewed the concept of noise shaping and explained the effect on the NTF, and consequently on the SQNR, of increasing the order of the modulator. Next, we will look briefly at the classification of modulators based on the types of signals they process, and describe some practical applications.

2.2 Classification of Delta-Sigma Modulators

Depending on the discretization of the time and amplitude axes of the input signal, modulators can be classified into the following three categories:

- Continuous-time (CT),
- Discrete-time, continuous-amplitude (DT analog),
- Discrete-time discrete-amplitude (DT digital).

2.2.1 Continuous-Time (CT) Analog Modulator

We consider two distinct implementations: sampled and unsampled quantizers.

2.2.1.1 Case 1: Sampled Quantizer: Synchronous Modulator

In this case, the input is a continuous-time analog signal and the modulator is implemented using continuous-time analog filters. The most common application is in continuous-time (CT) delta-sigma analog-to-digital converters (ADC) [6–21]. In a CT delta-sigma ADC, there is no need for an anti-aliasing filter or a front-end sampler. This simplifies system design by eliminating the anti-aliasing filter, which must precede other types of ADCs. In addition, the use of a CT filter postpones the inevitable sampling of the signal, which takes place at the output of the loop filter instead. Thus, imperfections in the sampling process arise at a much less sensitive point in the loop (see Fig. 2.6a) where the errors at this point are shaped by the NTF of the modulator [6].

The practical limit on the clock rate of a CT modulator is determined by the regeneration time of the quantizer and the update rate of the feedback DAC, whereas the clock rate in a discrete-time modulator is limited by the op-amp settling requirements [6]. In practice, a CT modulator can operate with a clock frequency which is 2–4 times greater than that which can be achieved with discrete time techniques [6].

2.2.1.2 Case 2: Unsampled Quantizer: Asynchronous Modulator

Asynchronous delta-sigma modulators can be used to convert an analog CT input signal into a CT discrete-amplitude output signal (see Fig. 2.6b). In this case, the information in the amplitude of the input signal is coded in the pulse widths of the output signal [22]. Due to its fully analog nature, the asynchronous delta-sigma modulator has a specific application area where pure analog processing is required, such as ADSL/VDSL line drivers, line drivers for optical cables, and UMTS transmitters [22]. In addition, asynchronous ADCs have been reported [23, 24].

2.2.2 Discrete-Time (DT) Analog Modulator

In the case of a discrete-time analog modulator, the input is a sampled analog signal and the modulator uses a discrete-time filter implemented with switched-capacitor

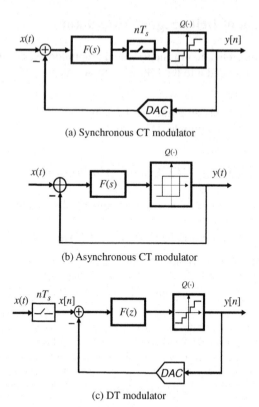

(a) Synchronous CT modulator

(b) Asynchronous CT modulator

(c) DT modulator

Fig. 2.6 Block diagrams of (**a**) a synchronous continuous-time modulator, (**b**) an asynchronous continuous-time modulator and (**c**) a discrete-time modulator

circuits. The main application of the DT analog delta-sigma modulator is in delta-sigma ADCs [6, 25, 26].

The CT and DT analog categories are beyond the scope of this book and the interested reader may refer to [6] for an excellent exposition of these types of modulators.

2.2.3 Discrete-Time Digital Modulators

In the case of a DT discrete-amplitude modulator, the input to the modulator is a quantized (digital) signal; consequently, the filters are implemented using a finite state machine. In this class, the delta-sigma modulator is implemented with digital circuits and we refer to this type of modulator as a digital delta-sigma modulator (DDSM). The main applications for this class of modulator are delta-sigma DACs [27–33] and delta-sigma fractional-N frequency synthesizers [34–42].

In DACs, the input can be a constant (DC) or a time-varying signal; in general, the input is time-varying. In fractional-N frequency synthesizers, the input to the DDSM is often a constant digital word. In this case, the DS-based frequency synthesizer is typically used as a local oscillator to generate channel frequencies in a transceiver. The synthesizer can also be used as a stand-alone transmitter that is capable of phase and frequency modulation; in this case, the input to the DDSM is a modulated data stream [43].

In this book, we are interested primarily in the case where the input to the DDSM is a constant digital word; this covers delta-sigma fractional-N synthesizers in the frequency generation application.

Before reviewing the application of DDSMs in DACs and fractional-N synthesizers, we first describe three common DDSM architectures [4, 6].

2.3 DDSM Architectures

Three types of modulators are investigated in this book: (1) single-quantizer DDSMs, (2) Error feedback modulators and (3) Multi stAge noise SHaping (MASH). In the following subsections, we will provide an overview of each of these architectures in turn.

2.3.1 Single Quantizer DDSMs

The general block diagram of a single-quantizer DDSM (SQ-DDSM) is shown in Fig. 2.7a. There is only one quantizer in the loop; hence the name "single quantizer" (SQ) DDSM. An example transfer characteristic of the digital multi-level quantizer Q is shown in Fig. 2.7b. The relationship between y and v is described by:

$$y = Q(v) = \begin{cases} R_{lo} + \frac{M}{2}, & v < R_{lo} \\ M \left\lfloor \frac{v}{M} + \frac{1}{2} \right\rfloor, & R_{lo} \leq v < R_{hi} \\ R_{hi} - \frac{M}{2}, & v \geq R_{hi}, \end{cases} \tag{2.21}$$

where $\lfloor x \rfloor$ denotes the largest integer less than or equal to x, M is a positive even integer referred to as the step-size, and $R_{lo} < R_{hi}$ are arbitrary odd multiples of $\frac{M}{2}$. For the characteristic shown in Fig. 2.7b, $R_{lo} = -\frac{5}{2}M$ and $R_{hi} = \frac{5}{2}M$.

The signal and the noise transfer function of the modulator in Fig. 2.7a are

$$\text{STF}(z) = \frac{F(z)}{1 + F(z)G(z)}, \tag{2.22}$$

and

$$\text{NTF}(z) = \frac{1}{1 + F(z)G(z)}, \tag{2.23}$$

respectively.

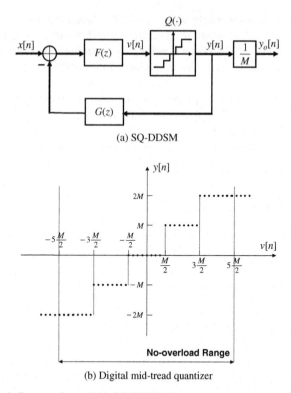

(a) SQ-DDSM

(b) Digital mid-tread quantizer

Fig. 2.7 (**a**) Block diagram of a multi-bit SQ-DDSM, (**b**) transfer characteristic of an example five level digital mid-tread quantizer with step size M

The filters characteristics $F(z)$ and $G(z)$ determine the signal and noise transfer functions and the order of the modulator. A special case is $F(z) = z^{-l}(1 - z^{-1})^{-l}$ and $G(z) = z^l - (z-1)^l$. These result in

$$\text{STF}(z) = z^{-l}, \tag{2.24}$$

and

$$\text{NTF}(z) = (1 - z^{-1})^l, \tag{2.25}$$

respectively.

Note that the STF is a delay of order l and the NTF is a high pass filter of order l. The squared magnitude of the frequency response of the NTF is $\left(2 \sin\left(\frac{\omega}{2}\right)\right)^{2l}$. At low frequencies ($\omega \approx 0$) this can be approximated by ω^{2l}. Therefore, the slope of the PSD is 60 dB/decade around $\omega \approx 0$ if $l = 3$.

In modulators of this type with order l greater than 1, the input range over which the quantizer is not overloaded[3] is a fraction of the full scale [4, 6, 44]. A third

---

[3] The stability of a DSM is often described in terms of the quantizer not being overloaded.

order modulator with the STF and NTF defined by Eqs. (2.24) and (2.25) can be overloaded if the quantizer has only one bit (i.e. two levels) [5]. By providing a larger number of quantizer levels, the input range over which the modulator is not overloaded increases. As an example, if $l = 3$ and $2^{l+1}+1$ output levels are allowed, then an input to the modulator which is less than half of the quantizer's full scale range is sufficiently small to prevent overload of the quantizer [6].

In order to maintain a higher order single-bit modulator in the no-overload region, the useful range of the delta-sigma modulator input signal must be reduced and more poles and zeros must be introduced within the feedback loop, compared to a multi-bit design with a comparable dynamic range [6, 45, pp. 23–33].

The classical SQ-DDSM is characterized by output feedback, i.e. the output y is fed back via the feedback network $G(z)$; an alternative architecture uses error feedback.

2.3.2 Error Feedback Modulators

In an Error Feedback Modulator (EFM), the quantization error is calculated and it, rather than the output, is fed back to the input via a filter H; that is why it is called *error* feedback. In fact, an EFM is a SQ-DDSM but it is often distinguished as a separate class.

In Fig. 2.8, we show the block diagram of a higher order EFM that uses a multi-bit quantizer (with a transfer characteristic such as that shown in Fig. 2.7b). Similar to the SQ-DDSM, a multi-bit quantizer is needed to maximize the dynamic range in a higher order modulator [6]. The filter H can have a general form, as in an SQ-DDSM. A special case that we consider is when $H(z) = 1 - (1 - z^{-1})^l$. This results in a NTF of the form $(1-z^{-1})^l$ like the SQ-DDSM discussed in Section 2.3.1.

2.3.3 MASH Topology

By contrast with the SQ-DDSM and EFM, both of which use a single quantizer, the multistage noise shaping (MASH) technique allows one to realize higher order noise shaping using lower order modulators. In the MASH DDSM, one can use lower order modulators (with orders as low as 1) in a cascade. If first order stages are available, l stages can be combined to form an lth order MASH modulator. For

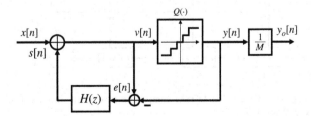

Fig. 2.8 Block diagram of an error feedback modulator

example, a third order MASH modulator can be implemented using three first order stages in cascade (denoted MASH 1-1-1). Alternatively, two stages can be used if one of the stages uses a second order SQ-DDSM (denoted MASH 1-2 or MASH 2-1). In the latter cases, the stability of the modulator is typically determined by that of the second order modulator.

A MASH DDSM with 1-bit internal first order modulators is a feedforward structure and is unconditionally stable [4]; this is the principal advantage of the MASH modulator over the SQ-DDSM topology. Furthermore, in a MASH DDSM, the stable input range is equal to the full scale while the stable input range is only a fraction of the full scale in an SQ-DDSM. The MASH DDSM is widely used in commercial fractional-N frequency synthesizer products.

Figure 2.9 shows the block diagram of a MASH DDSM where the stages in cascade use first order 1-bit error feedback modulators (denoted EFM1). The EFM1 block is shown in Fig. 2.10a and is equivalent to a first order delta-sigma modulator. The EFM1 modulator uses a 1-bit quantizer with the transfer characteristic shown in Fig. 2.10c. If the threshold point M is a power of 2, the EFM1 can be simply implemented using a conventional digital accumulator, as will be explained in detail in Chapter 3.

To obtain the lth order modulator in Fig. 2.9, l $EFM1$ stages are connected in cascade. The input x is applied to the first stage and the inverted quantization error of each stage is fed to the input of the following stage. The 1-bit outputs y_i of the stages are applied to a filter called the noise cancellation network. The function of the noise cancellation network is to eliminate the quantization noise contributions of all of the stages except the last one. In this way, the output contains information related to the input plus shaped quantization noise only from the last stage. The order of the noise shaping filter is equal to l. For example, consider the NTF and STF of a third order modulator. The linear model of the EFM1 shown in Fig. 2.10b is used to calculate the STF and NTF of the MASH DDSM. After straightforward calculations, one can show that $E(z) = -ME_q(z)$ where E_q is an additive error source with a non-zero mean $\left(-\frac{1}{2}\frac{M-1}{M}\right)$ in the range $\left\{0, -\frac{1}{M}, \cdots, -\frac{M-1}{M}\right\}$. For the three stages of the MASH 1-1-1 DDSM shown in Fig. 2.9 we write:

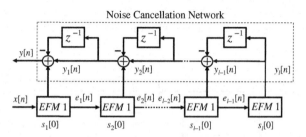

Fig. 2.9 Block diagram of a MASH DDSM comprising first order error feedback modulators (EFM1) of the type shown in Fig. 2.10a

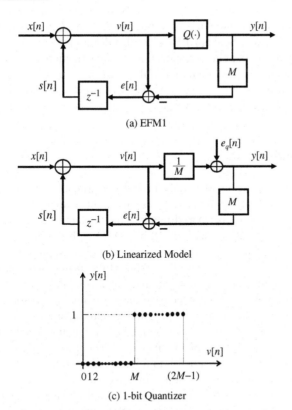

(a) EFM1

(b) Linearized Model

(c) 1-bit Quantizer

Fig. 2.10 Block diagrams of (**a**) a first order EFM (**b**) its linearized model and (**c**) transfer characteristic of the 1-bit quantizer used in EFM1

$$Y_1(z) = \frac{1}{M}X + (1 - z^{-1})E_{q1}(z) \tag{2.26}$$

$$Y_2(z) = \frac{1}{M}\left(-ME_{q1}(z)\right) + (1 - z^{-1})E_{q2}(z) \tag{2.27}$$

$$Y_3(z) = \frac{1}{M}\left(-ME_{q2}(z)\right) + (1 - z^{-1})E_{q3}(z), \tag{2.28}$$

where $Y_i(z)$ and $E_{qi}(z)$ are the z-transforms of the outputs y_i and the quantization errors, respectively.

The noise cancellation network multiplies the last equation by $(1 - z^{-1})^2$, the second equation by $(1 - z^{-1})$, the first equation by 1, and adds them together to produce:

$$Y(z) = Y_1(z) + (1 - z^{-1})Y_2(z) + (1 - z^{-1})^2Y_3(z) \tag{2.29}$$

$$= \frac{1}{M}X(z) + (1 - z^{-1})^3E_{q3}(z). \tag{2.30}$$

In this way, the quantization noise components E_{q1} and E_{q2} are cancelled exactly and the component E_{q3} is shaped by a third order high-pass filter[4] $(1 - z^{-1})^3$. MASH modulators are popular in applications such as fractional-N synthesis due to the simplicity of their implementation and their inherent stability [44].

Next, we consider the use of the DDSM in two application domains: digital-to-analog conversion and frequency synthesis.

2.4 Delta-Sigma DAC

In the following subsection, we study briefly the principles of operation of DS DACs [27–33] which are commercially as important as their ADC counterparts, and their implementation is often as difficult as the implementation of a delta-sigma ADC [6]. A high resolution (such as 18-bit) DAC with a relatively low voltage power supply is implemented effectively using the oversampling and noise shaping concepts. These techniques ease the design of the analog parts such that 18-bit accurate analog components are not required. Such resolution cannot be achieved in a conventional DAC without expensive analog trimming and/or an extremely long conversion time. The price one pays is the introduction of additional digital hardware operating at high frequencies (much higher than the signal bandwidth).

Figure 2.11a shows a typical block diagram of a delta-sigma DAC. Referring to Fig. 2.11, we will explain its operation in the frequency domain. The idea is not to convert the input bits to an analog signal directly, but to truncate the input digital word to a smaller number of bits. The truncated digital signal can then be converted to an analog signal with significantly less complex analog circuitry. The truncation is performed by the DDSM.

For simplicity, we have shown all the digital and analog spectra using the same x-axis, f. A bandlimited digital low pass signal u_1 with bandwidth $2f_B$, N_1-bit resolution and sampling frequency $f_{s1} = kf_B$ ($k \geq 2$) is applied to the system.

The spectrum of the high resolution digital signal u_1 contains the original baseband portion and its replicas located at integer multiples of f_{s1}, plus a small amount of quantization noise shown as a solid line in Fig. 2.11b. The interpolation filter (denoted IF), re-samples the digital input signal at a rate f_{clk} which is much greater than the original sampling frequency f_{s1} and applies the resulting oversampled signal to a digital low pass filter. The digital filter attenuates the images located between the baseband and f_{clk}.

The oversampled output signal u_2, with the resolution N_2-bits, is applied to the noise-shaping loop (NL), which is implemented as a DDSM. The images are attenuated by the digital filter in the IF block, relaxing the job of the CT reconstruction filter located after the sub-DAC. The DDSM truncates its N_2-bit input u_2 coarsely, resulting in u_3, which has a smaller number of bits (shown as m in the figure). The

---

[4] Note that the DC term due to non-zero-mean quantization noise is removed by the high pass filtering applied to the quantization noise.

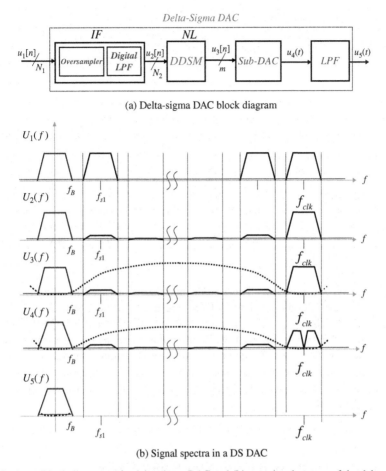

(a) Delta-sigma DAC block diagram

(b) Signal spectra in a DS DAC

Fig. 2.11 (**a**) Block diagrams of a delta-sigma DAC and (**b**) associated spectra of the delta-sigma DAC

quantization noise of the modulator is shaped so that most of its power is attenuated around the signal band, as illustrated by the dotted curve in the figure.

The fact that u_3 has only a few bits (at the minimum, only one bit) simplifies significantly the subsequent analog stages. u_3 is applied to an m-bit sub-DAC and the resulting analog output[5] u_4 is applied to an analog low pass reconstruction filter (LPF) to remove the quantization noise introduced by the DDSM, giving the desired analog output u_5.

After this short introduction, we report a few example specifications in Table 2.1. All have used multi-bit quantizers instead of a 1-bit quantizer for better stability

---

[5] We have assumed that the DAC transfer function with a zeroth order sample and hold is a sinc function defined by $\dfrac{\sin(\pi/(\mathrm{OSR}f_{s_1}))}{\pi/(\mathrm{OSR}f_{s_1})}$.

Table 2.1 Comparison of some delta-sigma DACs

Author	Hamasaki [27]	Adams [28]	Fujimori [30]	Annovazi [31]	Colonna [32]	Nguyen [33]	Lee [46]
Year	1996	1998	2000	2002	2005	2008	2009
Technology (CMOS μm)	0.6	0.6	0.5	0.35	0.13	0.18	0.35
DDSM order	3	2	3	3	3	2	3
DDSM type	NA	NA	SQ	SQ	SQ	NA	SQ
NTF type	NA	NA	Specific	Chebyshev	Specific	NA	Specific
Quantizer levels	5	64	31	13	17	256	7
SNDR (dB)	90	100	102	86	88	NA	69
DR (dB)	100	113	120	98	97	108	88
VDD (V)	3	5	5	3.3	3.3	1.8	0.8
Analog Power (mW)	10	NA	125	16.3	6.825	0.7	NA
Digital Power (mW)	12	NA	10	11.55	0.375	0.4	NA
Total power (mW)	22	125	155	27.85	7.25	1.1	1.3
Area (mm^2)	3.07	9.92	7.8	1.45	0.22	NA	1.76

NA indicates "Not Available".

and lower quantization noise.[6] Third order modulators are used in all cases except in [28, 33] where second order modulators are used. In these two cases, 64 and 256 quantizer levels were used to compensate for the reduction of SQNR with the lower modulator order. In most cases, the modulator is constructed using a single quantizer and the NTF is different from one design to another by having zeros at different frequencies. The typical dynamic range for audio applications is about 100 dB. One work [30] achieved 120 dB, but with significantly higher power consumption (155 mW [30]).

2.5 Phase-Locked Loop Frequency Synthesizers

The limited bandwidth available to each user in wireless systems mandates the precise definition of the carrier frequencies in both the transmit and receive paths. Frequency synthesizers generate periodic signals with accurately defined frequencies, thus serving as an integral part of radio frequency transceivers.

Frequency synthesis continues to be a challenge, fundamentally because performing algebraic operations on frequencies is more difficult than on other electrical quantities such as voltage or current. The challenge has taken different directions through the years, motivating the invention of various architectures and circuit techniques. As RF systems incorporate higher levels of integration, frequency synthesizers must deal with additional trade-offs resulting from application requirements such as monolithic implementation, low cost, minimal number of external components, and low power dissipation [47].

---

[6] The greater the number of quantizer levels, the smaller quantization error.

In the next subsection, we will first describe the basic idea of frequency synthesis using integer-N phase-locked loops. Then we will explain the principle of operation of a fractional-N synthesizer in order to obtain fine frequency resolution. We will illustrate how the performance of a fractional-N synthesizer can be improved with the aid of a DDSM.

2.5.1 Integer-N Frequency Synthesizers

Three architectures [44, 48] are commonly used to synthesize a desired frequency from a reference frequency: (i) table-look-up synthesizers, (ii) direct synthesizers and (iii) phase-locked loop (indirect) synthesizers. The first method cannot be used for high output frequencies, and the second approach is too bulky for silicon integration. Therefore, the phase-locked loop (PLL) is the dominant solution in frequency synthesizers [47] for wireless applications. In this case, the reference frequency is multiplied by a user-defined number. This is achieved by dividing the output frequency by that number, and adjusting the output frequency such that the divided frequency is equal to the reference frequency.

A PLL is a feedback system that operates on the excess phase of nominally periodic signals, i.e., the feedback operation in the loop automatically adjusts the phase of the locally generated signal $y(t)$ to match the phase of the fixed reference signal $x(t)$. As shown in Fig. 2.12, a PLL comprises a phase detector (PD), a low pass filter (LPF), and a voltage-controlled oscillator (VCO). The phase error between $x(t)$ and $y(t)$ is amplified and fed back so as to minimize the phase difference between $x(t)$ and $y(t)$. The loop is considered "locked" if the phase difference is constant; this corresponds to the input reference and output VCO frequencies being equal [49].

In the locked condition, the PLL operates as follows. The phase detector calculates the phase difference between the input reference and the VCO's output signal, and produces an output which is a function[7] of the phase difference. The low pass filter suppresses high frequency components from the PD output. The output of the filter is applied to the VCO to produce the desired output frequency. The VCO oscillates at a frequency that is equal to the input reference frequency but with a constant phase difference. In this way, the filter generates an appropriate control voltage for the VCO.

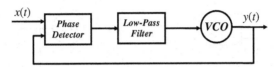

Fig. 2.12 A general phase-locked loop. VCO stands for voltage-controlled oscillator

---

[7] Ideally, the PD output is proportional to the phase difference.

Two PLL-based frequency synthesizer architectures are commonly used in applications today, namely integer-N and fractional-N synthesizers. The two implementations differ in how the divider is implemented and controlled. In this section, we discuss integer-N synthesizers. More details of fractional-N synthesizers will be given in later sections.

In Fig. 2.13, the PLL performs frequency multiplication, by means of a negative feedback path, to generate an output frequency f_{out} that is an integer multiple of the reference frequency f_{ref}. When the loop is locked,

$$f_{out} = f_{ref} \cdot N. \tag{2.31}$$

A reference frequency is provided to the phase detector for comparison with the divided VCO frequency f_{div}. In the "locked state", the VCO frequency is defined by Eq. (2.31). Programming the divider N with a new division number N can change the VCO output frequency, resulting in a frequency f_{out} that can be tuned across the overall band of interest. The primary constraint in this integer-N architecture is that the minimum channel spacing equals f_{ref}. As long as the loop is locked, the VCO output will have the same frequency resolution as the reference frequency, which is typically dependent on an external crystal oscillator. For example, with a reference frequency of 30 kHz and a division number of $N = 33000$, the VCO output frequency is 990 MHz. Assuming that the frequency accuracy of the oscillator is 1 ppm, the output of the VCO is accurate to ± 990 Hz around a 990 MHz carrier frequency, and the frequency resolution is 30 kHz.

The architectural simplicity of integer-N PLL frequency synthesizers has made them a popular choice for a variety of telecommunication systems [47]. However, the integer-N architecture has some significant drawbacks.

The frequency resolution, i.e. the channel spacing, is equal to the reference frequency, meaning that only integer multiples of the reference frequency can be synthesized. Therefore, if fine tuning is required, the designer's only choice in an integer-N PLL is to decrease the reference frequency.

Stability requirements limit the loop bandwidth to about one tenth of the reference frequency [45, 50]; therefore, decreasing the reference frequency increases the settling time as the loop bandwidth also has to be decreased. A large settling time is

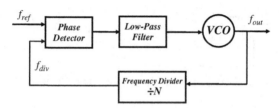

Fig. 2.13 Block diagram of a PLL-based integer-N frequency synthesizer. It comprises a phase detector, a low-pass filter and a voltage-controlled oscillator in the forward path and an integer frequency divider in the feedback path

not allowed by most communication standards [45]. Also, a reduced loop bandwidth allows less suppression of the VCO's inherent phase noise.

Another drawback of the integer-N PLL is the trade-off between phase noise and settling time when the divider ratio becomes large. The contributions to the output phase noise of almost all PLL building blocks, except the VCO, are multiplied by the division ratio [51]. A high output frequency resolution needs a small input reference frequency, which in turn requires a large divider value. A large divider value increases the inband noise, thereby increasing the rms phase error. In order to decrease the inband noise, the loop-bandwidth has to be kept low; this in turn increases the settling time.

In addition, if a small reference frequency is chosen, the reference spur[8] in the output phase noise is located at a smaller offset frequency. In order to suppress this spur, the loop bandwidth has to be decreased well below f_{ref}; this again increases the settling time.

In short, the design of integer-N PLL frequency synthesizers poses a trade-off between frequency resolution, spectral purity, and the PLL's dynamic behavior. An alternative way to obtain high resolution without compromising the dynamic performance is to implement fractional division. This is the topic of the next section, where we explain the principle of operation of a fractional-N frequency synthesizer.

2.5.2 Fractional-N Frequency Synthesizers

In fractional-N frequency synthesizers, fractional multiples of the reference frequency can be synthesized, allowing a higher reference frequency for a given frequency resolution. This in turn means that the loop bandwidth can be increased without compromising the spectral purity. Therefore, the PLL dynamics are accelerated and the total amount of capacitance required in the loop filter can be decreased so that single chip integration of the frequency synthesizer becomes feasible.

The basic idea behind fractional-N synthesis is division by fractional ratios, instead of only integer ratios [52, 53]. To accomplish fractional division, the same frequency divider is employed as in an integer-N frequency synthesizer, but the division is controlled differently. In Fig. 2.14, the division modulus of the frequency divider is controlled by the carry output of a simple digital accumulator[9] that is n_0-bits wide. To realize a fractional division ratio $N = N_0 + \beta$, with $\beta \in \left\{0, \frac{1}{2^{n_0}}, \frac{2}{2^{n_0}}, \cdots, \frac{2^{n_0}-1}{2^{n_0}}\right\}$, a digital input $X = \beta \cdot 2^{n_0}$ is applied to the accumulator. With the carry output c as the control signal, X ones and $2^{n_0} - X$ zeros are generated for every 2^{n_0} output samples. When the carry out is one, the divider value is set to $N_0 + 1$, and the divider value is set to N_0 when the carry out is zero.

---

[8] The reference spur refers to unwanted frequency modulation of the VCO at the reference frequency, f_{ref}.

[9] As we will see in Chapter 3, the accumulator can be considered as a first order DDSM.

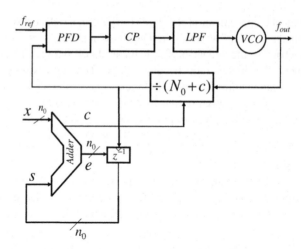

Fig. 2.14 Block diagram of a fractional-N PLL with a digital accumulator controlling the division ratio. The synthesizer includes a phase frequency detector (PDF), LPF, a VCO, and a variable modulus divider. The carry out of the adder c controls the divider modulus

This means that the frequency divider divides $2^{n_0} - X$ times by N_0 and X times by $N_0 + 1$, resulting in a division ratio N_{mean}, given by:

$$N_{mean} = \frac{(2^{n_0} - X) \cdot N_0 + X \cdot (N_0 + 1)}{2^{n_0}} \qquad (2.32)$$

$$= N_0 + \frac{X}{2^{n_0}} = N_0 + \beta.$$

Therefore,

$$f_{out} = (N_0 + \beta) f_{ref}. \qquad (2.33)$$

In this case, the frequency resolution Δf is defined by

$$\Delta f = \frac{1}{2^{n_0}} f_{ref}. \qquad (2.34)$$

Equation (2.34) shows that, for a given reference frequency, it is possible to make the frequency resolution arbitrarily high by making the accumulator wordlength sufficiently large. For example, in a DCS-1800 telecommunication system, the channel spacing of 200 kHz can be accommodated by selecting $f_{ref} = 26$ MHz and using an accumulator width n_0 of more than 8 bits.

While dynamically switching the divider modulus solves the problem of achieving non-integer multiples of the reference frequency, a price is paid in the form of increased phase noise resulting from jitter in the feedback signal. During each reference period, the difference between the actual modulus (N_0 or $N_0 + 1$) and the desired average modulus $(N_0 + \beta)$ represents a phase error. This error gets injected into the PLL and results in increased phase noise. The amount by which the phase noise is increased depends upon the characteristics of the output of the divider moduli or, equivalently, the spectrum of the signal that controls the divider in the feedback loop.

2.5.3 Spurious Tones

In the digital accumulator implementation of a fractional-N synthesizer shown in Fig. 2.14, once the wordlength n_0 and input X are fixed, the carry out control signal exhibits periodic behavior. In fact, since the accumulator is a finite state machine, it cycles through its states in a periodic manner, the length of the cycle depending on X and n_0. As an example, let us assume that the reference frequency is 1 MHz, $X = 5$, $n_0 = 3$ and $N = 100$. With these values, the output frequency is calculated from Eq. (2.33) as

$$f_{out} = \left(100 + \frac{5}{2^3} \right) 1 = 100.625 \text{ MHz.}$$

The signal values of s in the accumulator for 11 cycles of the reference clock are summarized in Fig. 2.15. We assume that the initial state $s[0]$ of the register is zero. The carry out signal c is determined by:

$$c = \begin{cases} 0, & X + s < 8 \\ 1, & X + s \geq 8, \end{cases} \tag{2.35}$$

and the error e is equal to $(X + s)$ modulo 8.

Clock count	X	s	e	X+s	c
0	5	0	5	5	0
1	5	5	2	10	1
2	5	2	7	7	0
3	5	7	4	12	1
4	5	4	1	9	1
5	5	1	6	6	0
6	5	6	3	11	1
7	5	3	0	8	1
8	5	0	5	5	0
9	5	5	2	10	1
10	5	2	7	7	0

Fig. 2.15 The first 11 samples of the signal values in a three bit accumulator with input $X = 5$ and $s[0] = 0$

Note that the internal state s and the carry out signal c are periodic with a period of 8 clock cycles. When $X = 5$, five ones are generated during each period so that the average value of the carry out signal over a period is equal to $\frac{5}{8}$, corresponding to the desired fraction. This enables one to generate the output frequency of 100.625 MHz and steps of $\frac{1}{8}f_{ref} = 0.125\,\text{MHz}$. Each time the carry out signal is unity, it sets the divider value to 101. The fact that the actual divider value (100 or 101) is different from the desired divider modulus (100.625) shows up as an instantaneous phase difference at the input to the phase frequency detector. This phase error is determined by the nature of the carry out signal and causes a phase error (which is periodic in this case). The periodicity in the phase error results in a tonal spectrum. Any tones that are located outside the PLL bandwidth are attenuated by the LPF. However, those that are inside the PLL's bandwidth pass through the LPF and modulate the VCO frequency, manifesting themselves as undesirable spurious tones (so-called "spurs") in the output phase noise.

These tones are also called fractional tones because they are located at fractional multiples of the reference frequency. In our example, the period is 8 reference cycles; therefore, the first tone is located at $\frac{1}{8}f_{ref} = 0.125\,\text{MHz}$, which can be inside the loop bandwidth. Such spurs are not tolerated by most wireless communication standards [44, 47].

One way to attenuate these tones is to decrease the loop bandwidth. However, this solution negates the principal advantage of the fractional-N synthesizer, namely that of having fine tuning resolution while maintaining a relatively large loop bandwidth.

Another way to eliminate fractional tones is to introduce randomness to break the periodicity in the sequence of the division moduli while still achieving the desired average modulus [45]. One can generate a control sequence that approximates a sequence of independent random variables that take on the values 0 and 1 with probabilities of $1 - \frac{X}{2^{n_0}}$ and $\frac{X}{2^{n_0}}$, respectively. During the nth reference period, the divider modulus is still N_0 or $N_0 + 1$ with the prescribed probabilities. However, the resulting sequence of moduli has the desired average but the power spectral density of the error is spread uniformly over many frequencies. In this way, fractional tones can be eliminated. Instead of tones, this modified technique ideally introduces white noise with a low PSD. Unfortunately, the portion of the white noise within the PLL's bandwidth is integrated by the PLL. Consequently, the overall contribution to the phase noise can be significant unless the PLL bandwidth is small.

Alternatively, one can generate a randomized control sequence whose power at frequencies below the loop bandwidth is highly attenuated [54–56]. A DDSM can generate such a noise shaped control sequence whose power is located mostly outside the PLL loop bandwidth.

An example of a delta-sigma fractional-N PLL is shown in Fig. 2.16. The system comprises a phase frequency detector, a charge pump, a LPF and a VCO in the forward path and a controlled multi-modulus divider in the feedback path. Like the simple accumulator implementation, the DDSM output controls the divider moduli in order to implement fractional division. The input to the DDSM is a constant digital value X and it is clocked by the output of the divider. The DC component of the DDSM's output power spectrum is proportional to X; therefore, the time average

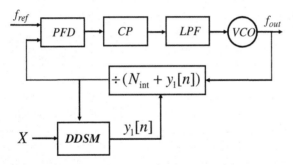

Fig. 2.16 Block diagram of a delta-sigma fractional-N PLL

of the output sequence y_1 is proportional to X; this sets the desired fraction. The divider divides the VCO output frequency f_{out} by the integer $N_{int} + y_1[n]$, where N_{int} is a fixed integer and $y_1[n]$ is the DDSM output at each instant n. Over many reference cycles, the average of $N_{int} + y_1[n]$ approaches $N_{int} + \beta$, implementing (in a time-average sense) the fractional division β.

In short, the DDSM with input X is used to perform the following operations:

- Its output signal y_1 gives the desired fraction β on average.
- y_1 possesses a colored spectrum. Firstly, it has a DC tone whose amplitude is proportional to X, setting the desired fraction.[10] Secondly, the low frequency part of the rest of its spectrum is attenuated and its high frequency part is amplified. The former is desirable because it rejects the portion of the power spectrum that passes through the LPF. The latter is undesirable; however, it is rejected in principle by the LPF.
- A higher order DDSM[11] randomizes the error sequence with or without one of the auxiliary randomization techniques described in Chapters 3, 4 and 5, to break patterns resulting from short periodic cycles.

In the following, we show by simulation how the delta-sigma modulator can be used to remove fractional tones. Firstly, in Fig. 2.17, we show for comparison the PSDs of the output signals of a digital accumulator and a third order MASH delta-sigma modulator of the type shown in Fig. 2.9. For both configurations, the wordlength is $n_0 = 18$ and the input is $X = 256$. The initial value of the register in the accumulator was set to unity. The initial values of the MASH DDSM were set to 1, 0 and 0, for the first, second and third stages, respectively. As we will see in Chapter 3, the period of the output of the accumulator for this combination of input

---

[10] Note that the DC term due to non-zero-mean quantization noise is removed by the high pass filtering applied to the quantization noise.

[11] The simple accumulator is a first order DDSM; it fails to perform proper noise shaping and randomization, as we will see in Chapter 3. A higher order modulator is required for adequate randomization of the quantization noise.

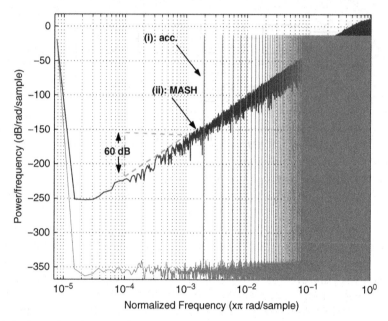

Fig. 2.17 Spectral plots of the output signals of simple accumulator [plot (*i*)] and, a third order MASH DDSM [plot (*ii*)], both with $n_0 = 18$, $X = 256$. The envelope of the third order DDSM has a slope of 60 dB/decade at low frequencies

and initial conditions is $\frac{2^{18}}{2^8} = 1024$. Therefore, we would expect that the first (fundamental) tone in the accumulator output spectrum should be located at $\frac{2\pi}{1024}$. Plot (i) in Fig. 2.17 confirms that the first tone is located at approximately $2 \times 10^{-3}\pi$, as expected. By contrast, the third order MASH DDSM randomizes the output control sequence and the power of its quantization noise is distributed over many tones so that the output spectrum is smoothly shaped toward higher frequencies, as shown in Fig. 2.17.

Note that we have plotted the spectra using logarithmic axes, whereas we used linear axes in the introduction to the delta-sigma modulation at the beginning of this chapter. Logarithmic axes are mainly used in the literature for characterizing noise shaping in a delta-sigma modulator. In this example, we have a third order modulator; consequently, the PSD is proportional to ω^6 at lower frequencies. Hence, the slope is 60 dB/decade, as can be seen from the figure.

In this example, the accumulator and the MASH DDSM were implemented in a Verilog-AMS behavioral model of the PLL [57]. The modulators were modelled using Verilog. The PFD, CP, VCO and the divider were described by behavioral Verilog-AMS models. The third order loop filter was implemented based on passive capacitors and resistors. The VCO and the divider blocks were merged into one block in order to decrease the simulation time [51]. The loop parameters are as follow: the reference frequency is 26 MHz; the loop bandwidth is 300 kHz; the LPF order is three; the input to the modulators is 256; the wordlength of the accumulator and the MASH DDSM is 18; the integer divider is 68; the VCO gain is 70 MHz/V and the charge pump current is 1 mA. For these simulations, we set the up and down currents of the CP to be equal and minimize the noise sources and jitter of the VCO, PFD, CP, reference input and digital blocks to enable us to examine the quantization noise of the modulators in isolation.

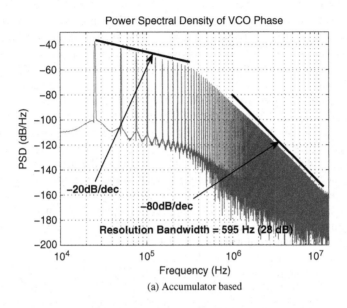

(a) Accumulator based

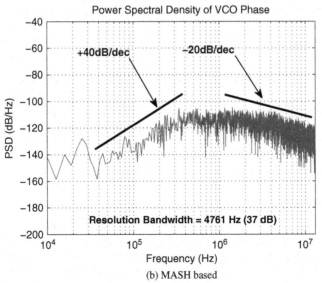

(b) MASH based

Fig. 2.18 Phase noise plots of (**a**) accumulator and (**b**) MASH based fractional-N PLLs

In Fig. 2.18a, b, we show the resulting phase noise plots. The period of the accumulator output is 1024 reference cycles; therefore, the first fractional tone in the phase noise plot is located at an offset of $\frac{f_{ref}}{1024} \approx 25.4$ kHz. This fractional tone, and tones close to it, lie within the loop bandwidth of the PLL; therefore, they show up in the phase noise plot in Fig. 2.18a with higher power than those located outside the PLL bandwidth. The third order MASH DDSM fixes the problem of fractional tones, as can be seen in Fig. 2.18b. None of the high power fractional tones in the spectrum of the accumulator-based PLL is present in the case of the MASH-based PLL. However, since most of the quantization noise produced by the DDSM is pushed toward higher frequencies, the out-of-band phase noise content is larger than for the accumulator based PLL, falling off at -20 dB/decade instead of -80 dB/decade.

In this example, the MASH DDSM has been configured in such a way that it exhibits noise shaping without producing fractional tones. In general, however, most DDSM configurations may have reduced performance due to the nature of their digital implementation. In fact, if the modulators are not designed properly or special measures are not taken, they can exhibit qualitatively similar performance to that of the simple accumulator, namely having fractional tones, otherwise called "spurious tones" or spurs.

The accumulated quantization noise appears as a phase error at the PFD input. This accumulation process contributes to a -20 dB/decade slope in the phase noise plots. As shown in plot (i) of Fig. 2.17, the slope of the PSD of the accumulator is 0. Due to the prescribed phase conversion, the slope in Fig. 2.18a below the loop bandwidth is -20 dB/decade and above the loop bandwidth is $-20 - 60 = -80$ dB/decade. The -60 dB/decade term is due to the roll-off of the third order loop filter. By contrast, in the case of the MASH-based PLL, the slope below the loop bandwidth is $+60 - 20$ dB $= +40$ dB/decade and the slope is $+60 - 20 - 60 = -20$ dB/decade above the loop bandwidth.

2.6 Simulink Models and MATLAB Codes for DDSMs

In this section, we provide Simulink models and MATLAB code[12] for three sample DDSM architectures, namely SQ, EFM and MASH. First we study the SQ-DDSM.

2.6.1 SQ-DDSM

2.6.1.1 Simulink

Figure 2.19 shows the block diagram of a third order SQ-DDSM [9]. This modulator implements the STF and NTF given in Eqs. (2.24) and (2.25) with $l = 3$, namely $STF(z) = z^{-3}$ and $NTF(z) = (1 - z^{-1})^3$. These can be found by replacing

---

[12] All MATLAB and Simulink files described in this book are available for download from: http://cas.tyndall.ie

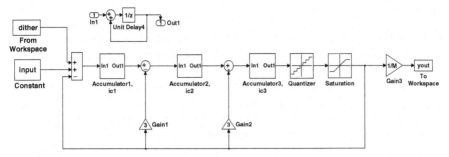

Fig. 2.19 Block diagram of a third order SQ-DDSM (SQDDSM3rd.mdl) simulated using Simulink

the quantizer and saturation block with a gain factor ($k = 1$) and an additive noise source e_q and finding the output of the saturation block in the z domain as a function of the main input and e_q.

The main input to the modulator is a constant value. A pseudorandom binary "dither" sequence generated in MATLAB, taken from the workspace, is added to the constant input to break up periodic cycles, as we will discuss in detail in the next chapter. There are three identical accumulators in the loop. The details of the first accumulator are shown in the figure. The initial value of the internal unit delay is set to ic_i in Accumulator$i$. The Quantizer and Saturation blocks implement a mid-tread digital quantizer like the one shown in Fig. 2.7b.

In Table 2.2, we show how each block is configured and in which library it can be found.

The Start time and the Stop time in the Simulation Parameters menu are set to 1 and sim_time. The type in the Solver option is set to "Fixed-step" and "discrete (no continuous step)". This setup assumes that the time index is unit-less integer $(1, 2, 3, \cdots,$ sim_time$)$ and that the signals are signed integers.

The model is saved as a ".mdl" file to be used in the MATLAB code. With the set of parameters described, the Simulink model is ready for simulations using the MATLAB code given in the next subsection.

Table 2.2 Details of the blocks in Fig. 2.19

Block	Library	Configuration
Constant	Sources	Constant value=input
From workspace	Sources	Data=dither
Quantizer	Discontinuities	Quantization interval=M
Saturation	Discontinuities	Lower limit=n_min$\times M$; upper limit=n_max$\times M$
To workspace	Sinks	Variable name=yout, save format=array
Gain	Math operations	
Unit delay	Discrete	Initial condition=ic$i$ for accumulator i
Sum	Math operations	

2.6.1.2 MATLAB

In this subsection, we describe the MATLAB code that runs the Simulink model described in the previous subsection and generates the power spectral density of the output of the DDSM. The quantizer step size is $M = 2^{16}$; the input is $\frac{M}{2}$, sim_time $= 2^{18}$, n_min $= -4$ and n_max $= 4$. The output of the DDSM takes on values in the range $\{-4, -3, \cdots, 3, 4\}$. The generated "dither" signal can be selected using the d_sw flag. The dither signal is used in the Simulink model as shown in Fig. 2.19 and it is generated by the following operation:

$$round(rand(1, simtime)).$$

The dither is selected and the initial conditions are set to zero. The Simulink model SQDDSM3rd.mdl is then simulated using the command "sim". After simulation, the average value of the output is calculated as a quick check for correct operation.

The average in this case should be equal to $\frac{\frac{M}{2}}{M} = 0.5$.

```
% Define the step size
n=16;
M=2^n;
% Define the max and min values of the saturation block
n_min=-4;
n_max=4;
% Define the number of simulation points
sim_time=2^18;
% Define the input
input=1*M/(2^1);
% Calculate the variance for linear prediction
variance=1/12;
% Enable (d_sw=1) or disable (d_sw=0) dither
d_sw==1;
if d_sw=1
dither=[1:sim_time;1*round(rand(1,sim_time))]';
else
dither=[1:sim_time;zeros(1,sim_time)]';
end
% Set the initial condition
ic1=0;
ic2=0;
ic3=0;
% Simulate the Simulink model for the third order SQDDSM
sim('SQDDSM3rd.mdl',[1 sim_time]);
% Check the output average
m_y=mean(yout);
% Define a Hanning window
```

```
window=hann(length(yout));
% Calculate the PSD using Periodogram
[Pyy, w] = periodogram(yout,window,sim_time);
% Consider dither and calculate the PSD using the linear prediction
dither_contribution=1/12*1/(M^2);
PSD_predicted=dither_contribution+variance*(2*sin(w/2)).^6;
% Plot output PSD and the PSD using the linear prediction
figure;
semilogx(w/pi,10*log10(pi*Pyy),'b');
hold
semilogx(w/pi,10*log10(PSD_predicted),'k');
grid on
xlabel('Normalized Frequency (x π rad/sample)','fontsize',14)
ylabel('Power/frequency(dB/rad/sample)','fontsize',14);
set(gca,'fontsize',14)
set(findobj(gca,'Type','line'),'LineWidth',2)
```

The PSD of the output is calculated using the "Periodogram" [7] that calculates:

$$S[k] = \frac{\frac{1}{N_f}\left|\sum_{l=0}^{N_f-1} w[l]x[l]e^{-j\frac{2\pi}{N_f}kl}\right|^2}{\frac{1}{N_f}\sum_{l=0}^{N_f-1}|w[l]|^2}, \quad 0 \le k < N_f \qquad (2.36)$$

where $x[l]$ is the lth sample of the signal x with length N_f, $w[l]$ is the lth sample of the window signal w, and k is an integer in the given range. A Hanning window (window=hann(length(yout));) with the same length of the signal is used; it is generated in MATLAB using the command "hann" that calculates the following equation:

$$w[k] = 0.5\left(1 - \cos\left(2\pi\frac{k}{N_w - 1}\right)\right), \quad 0 \le k \le N_w - 1 \qquad (2.37)$$

where N_w is the length of the window.[13] All the spectral plots in this book use the Periodogram function with a Hanning window, unless otherwise stated.

Considering the dither contribution, the white noise prediction is calculated using

$$S(\omega_k) = \frac{1}{M^2}\frac{1}{12} + \sigma^2\left\{2\sin\left(\frac{\omega_k}{2}\right)\right\}^{2l}, \quad \omega_k \in \left\{0, \frac{2\pi}{N_f}, \frac{4\pi}{N_f}, \cdots, \pi - \frac{2\pi}{N_f}\right\}, \qquad (2.38)$$

where $\sigma^2 = \frac{1}{12}$ and l is the order of the modulator; $l = 3$ in this example.

---

[13] Here N_w =sim_time.

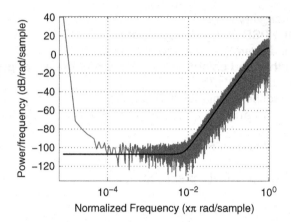

Fig. 2.20 Simulation result after running the MATLAB code. The DC component corresponds to $X = \frac{M}{2}$

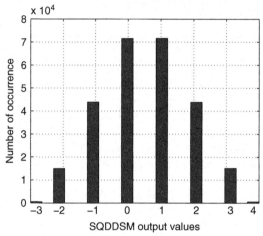

Fig. 2.21 Distribution of the output samples for the simulated third order SQ-DDSM

Figure 2.20 shows the simulation result after running the prescribed MATLAB code. The noise floor is due to the dither contribution and, as one can see, the quantization noise is pushed away from lower frequencies, as expected.

In Fig. 2.21, we show the distribution of the output samples and in Fig. 2.22 we show some output samples for illustrative purposes. The output in this simulation example occupies 8 levels, as shown in the histogram plot.

2.6.2 Multi-Level EFM

We consider a third order modulator with an all pass STF ($STF(z) = 1$) and an NTF given by Eq. (2.25) with $l = 3$. The Simulink model is shown in Fig. 2.23.

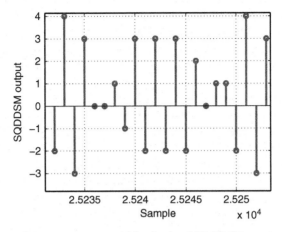

Fig. 2.22 A portion of the output sequence of the simulated SQ-DDSM

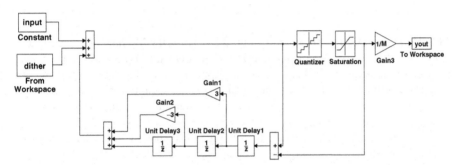

Fig. 2.23 Simulink model of a third order EFM DDSM (EFM3rd.mdl)

The configuration parameters and the MATLAB code are identical to those of the SQ-DDSM in Section 2.6.1. To calculate the STF and NTF of the modulator, the quantizer and the saturation blocks are replaced by a gain factor ($k = 1$ in this case) and an additive noise source. After simple algebric calculations in the z domain one can determine the STF and NTF.

2.6.3 MASH

In this subsection we consider the MASH DDSM. First we describe the Simulink model and then we explain the MATLAB code.

2.6.3.1 Simulink Model

The MASH DDSM architecture we consider in this book is based on the first order error feedback modulator shown in Fig. 2.10a. The Simulink model for the popular

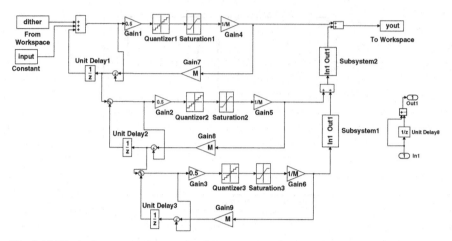

Fig. 2.24 Block diagram of the simulated third order MASH 1-1-1 DDSM (MASH111.mdl) in Simulink

MASH 1-1-1 DDSM is presented in Fig. 2.24. This DDSM contains three identical EFM1 stages. The first stage comprises Gain1, Quantizer1, Saturation1, Gain4, Gain7, a subtractor, Unit delay1 and the input summing block. The gain block (Gain1) with the value of 0.5, the quantizer block, the saturation block and the gain block of $\frac{1}{M}$ are explicitly used in order to implement the 1-bit quantizer described in Fig. 2.10c. The noise cancellation network comprises Subsystem1, Subsystem2 and two summing blocks. The details of the identical subsystem blocks are shown next to Subsystem1. The configuration parameters of the blocks in the MASH 1-1-1 Simulink model are summarized in Table 2.3.

2.6.3.2 MATLAB Code

The MATLAB code is very similar to that presented in Section 2.6.1.2. The output of each EFM1 stage in the MASH modulator is binary (0 or 1). After applying these three 1-bit outputs to the noise cancellation network, the final output of the modulator contains 8 levels in the range $\{-3, -2, \cdots, 3, 4\}$. A representative output

Table 2.3 Details of the blocks in Fig. 2.24

Block	Library	Configuration
Constant	Sources	Constant value=input
From workspace	Sources	Data=dither
Quantizer	Discontinuities	Quantization interval=M
Saturation	Discontinuities	Lower limit=0; upper limit=M
To workspace	Sinks	Variable name=yout, save format=array
Gain	Math operations	Values shown in Fig. 2.24
Unit delay	Discrete	Initial condition=ic1, ic2 and ic3 for unit delays 1, 2 and 3 respectively
Sum	Math operations	

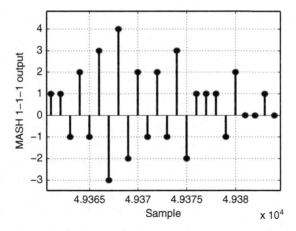

Fig. 2.25 A portion of the output sequence of the MASH DDSM with the constant input $\frac{M}{2}$ where $M = 2^{16}$

sequence is shown in Fig. 2.25. The average of the output sequence converges to the input divided by M (0.5 in the case that $X = \frac{M}{2}$). The designer needs first to check the average of the output sequence to verify correct operation. The second step is to observe the PSD. When checking the PSD, the designer should measure the slope of the quantization noise spectrum. In the case of the MASH 1-1-1 DDSM, the slope is 60 dB/decade. In Fig. 2.26 we show the output PSD of the MASH 1-1-1 DDSM when dithered with LSB dithering signal at input. As shown in the figure, the slope is 60 dB/decade and there are no obvious spurs in the spectrum.

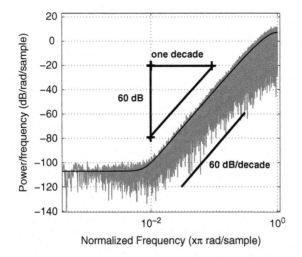

Fig. 2.26 Output PSD of the simulated MASH 1-1-1 DDSM with dither

2.7 Summary

In this chapter, we have reviewed in detail the idea of quantization noise shaping and have described how delta-sigma modulation may be used to implement this concept. In principle, the inband signal-to-quantization noise ratio at the output of a DDSM can be improved by using higher order modulators and by increasing the oversampling ratio.

We classified modulators based on their input signals. Analog modulators belong to two categories: discrete-time and continuous time. They are used primarily in analog-to-digital converters. DDSMs are used in digital-to-analog converters and fractional-N frequency synthesizers. In this book, we are interested primarily in applications where the input to the DDSM is a constant digital word.

We reviewed representative DDSM topologies including single quantizer, error feedback and MASH modulators.

The principles of operation of the delta-sigma DAC were reviewed briefly. Following an explanation of the DAC, the basics of fractional-N frequency synthesizers were presented. We explained the motivation behind the use of a DDSM for implementing fractional frequency division. Through the simulation of a PLL, we showed that the phase noise of a fractional-N synthesizer can be improved if a DDSM is used instead of a simple accumulator for controlling the divider modulus. The idealized DDSM randomizes and noise shapes the divider sequence, thereby removing the fractional tones that appear in the case of an accumulator based synthesizer.

We concluded the chapter by providing Simulink models, MATLAB code and representative simulations for three DDSMs, namely SQ, EFM and MASH.

Chapter 3
Conventional Techniques for Maximizing Cycle Lengths

3.1 Introduction

In Chapters 1 and 2, we introduced the problem of short cycle lengths in DDSMs. We observed that the spectrum contains undesirable tones when the period of the quantization error signal is short. In this chapter, we review conventional techniques that guarantee long cycles.

There are two classes of techniques for maximizing cycle lengths in DDSMs: "stochastic" and deterministic. The "stochastic" approach to maximizing cycle lengths is to use a "random"[1] dither sequence to disrupt periodic cycles [4, 8–10]. Dithering breaks up the cycles and increases the effective cycle length, resulting in smooth noise-shaped spectra. While the "stochastic" solution increases the cycle length, as required, it inherently adds noise to the spectrum; care must be taken to minimize the contribution of this additional noise. By contrast, the deterministic approach to whitening the quantization noise is to guarantee maximum cycle lengths by design, without explicitly adding noise.

In the deterministic category, one approach is to avoid known short cycles by setting the initial conditions of the internal registers of the constituent stages. Recent work [5, 11, 13] has addressed this technique. Seeding the first stage with an odd value in the MASH DDSM has been shown, by empirical observations [5, 11] and mathematically [13, 58], to maximize the cycle length. A second way of maximizing the cycle length in a deterministic manner is to use a prime modulus quantizer [15].

In this chapter, we explain stochastic and deterministic techniques in detail. First, we will describe stochastic techniques. Dithering belongs to this category and it is the most popular randomization method in the design of DS modulators. Secondly, we will describe two deterministic techniques that maximize cycle lengths by design: (a) setting predefined initial conditions (seeding) and (b) choosing a prime modulus for the quantizer.

The seeding technique has been examined in [11] using extensive computer simulations. We will describe in detail observations from that work based on simulations and we will present a mathematical proof [13] of the empirical results reported

---

[1] In practice, the "random" sequence might be generated by a pseudo-random or chaotic source.

K. Hosseini, M.P. Kennedy, *Minimizing Spurious Tones in Digital Delta-Sigma Modulators*, Analog Circuits and Signal Processing, DOI 10.1007/978-1-4614-0094-3_3, © Springer Science+Business Media, LLC 2011

in [11] for the case of a MASH topology with single bit quantizers. We will derive an exact expression for the cycle length in the MASH 1-1-1 as a function of the (constant) input and initial conditions [14].

In addition to the seeding technique, we will describe another method that uses prime modulus quantizers [15] instead of conventional power-of-two modulus quantizers. We show mathematically that, in the case of an lth order MASH modulator [16], this technique guarantees a minimum cycle length that is equal to a prime number.

3.2 Stochastic Techniques

As mentioned above, there are two classes of techniques for maximizing the cycle length in a DDSM. In the second part of this chapter, we will study two deterministic approaches in detail, but first we review stochastic methods. In the so-called dithering approach, a *random* dither signal is injected into the system in order to disrupt short cycles. The random signal can be generated by a chaotic source, or, more typically, a pseudorandom binary sequence generator (PRBS).

Dithering is the method that is most commonly used in practice to improve a modulator's spectral performance [4]. Ideally, dithering improves the statistical properties of the quantization noise such that the white noise approximation applies. If the error introduced by the quantizer[2] is white, the shaped output quantization noise should be free of spurious tones.

In the following subsections, we describe some practical dithering topologies.

3.2.1 Nonshaped LSB Dithering

The block diagram shown in Fig. 3.1 illustrates the additive nonshaped input dithering technique [9, 59–61]. The term nonshaped refers to the fact that the dither signal is added to the input without filtering. A 1-bit dither signal d is added directly to the LSB of the input x of the modulator; hence it is called additive input LSB dithering.

The dither signal is assumed to be uncorrelated with the input x and therefore its white power spectrum is added directly to the spectrum of the input signal. The total spectrum applied to the modulator is filtered by its signal transfer function

Fig. 3.1 Block diagram of a DDSM with nonshaped LSB dithering

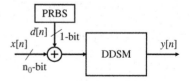

---

[2] Here, we refer to the quantization error of a quantizer in the DDSM that contributes to the output quantization noise.

(STF). If the STF has a low pass or an all pass character, the modulator passes the low frequency portion of the input spectrum, including the dither's spectrum. In this way, nonshaped LSB dithering degrades the in-band content of the output spectrum by raising the noise floor of the output signal. This effect is particularly significant when the number of input bits n_0 is low because the dither power can be comparable to the input power. In other words, the ratio of the amplitudes of the dither signal and the input signal should be kept low enough so that the in-band signal-to-quantization noise ratio of the output signal is not seriously degraded.

One way to decrease the noise floor is to increase the wordlength of the modulator; this is illustrated in Fig. 3.2. By increasing the input word length from 10 to 18 bits in a MASH 1-1-1 modulator (see Fig. 3.13), the noise floor decreases by approximately 48 dB.

Pamarti et al. have studied the effectiveness of LSB dithering in two classes of modulators: (a) the single quantizer DDSM (SQ-DDSM) [61], and (b), the MASH DDSM [9]. We present their results next.

3.2.2 LSB Dithering in SQ-DDSMs

In the case of the SQ-DDSM, Pamarti et al. [61] have developed necessary and sufficient conditions, mainly on the impulse response of the forward filter in the

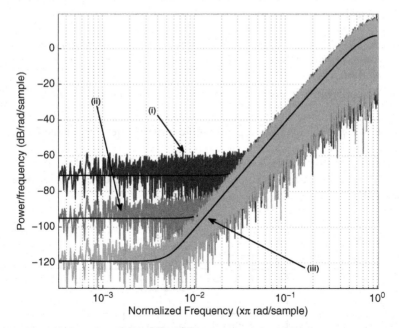

Fig. 3.2 Spectral plots of a MASH 1-1-1 DDSM with nonshaped LSB dithering at the input for three cases of input word length n_0: (*i*) 10, (*ii*) 14, and (*iii*) 18. The input is $X = \frac{M}{4} + \frac{M}{8}$, where $M = 2^{n_0}$. The *solid curves* represent the shaped white quantization noise, including the dither contributions where applied. By increasing n_0, the noise floor decreases (by 6 dB/bit)

DDSM, to ensure that the quantization noise of the nonoverloaded quantizer in the DDSM is asymptotically uniformly distributed and independent of delayed versions of itself and the input [61].

Theorem 1 is divided in two sections: (a) and (b). A non-overloaded SQ-DDSM shown in Fig. 3.3 with a multi-level quantizer having a transfer characteristic like that shown in Fig. 2.7b is assumed, where x is the desired bounded input, and d is the 1-bit dither signal added to x.

U and V are independent integer random variables that are uniformly distributed over $[-\frac{M}{2} + 1, \frac{M}{2}]$ with a mean of $\frac{1}{2}$ and variance $\frac{M^2-1}{12}$. The two sections of the theorem are:

(a) The quantizer error $e_q = y - v$, is *asymptotically identically distributed and independent* of x_i if and only if at least one of the following conditions is true for each positive integer p and for each integer $k, 0 < k < M$:

  • $(kf[n] \bmod M)$, where f is the impulse response of the forward filter $F(z)$ shown in Fig. 3.3, does not converge to zero as $n \to \infty$.
  • A non-negative integer $s = s(p) \neq p$ exists such that $kf[s] \bmod M = \frac{M}{2}$.

(b) The pair $(e_q[n], e_q[n - p])$ converges in distribution to the prescribed pair of independent random variables (U, V) if and only if at least one of the following conditions is true for each positive integer p, and each integer pair $(k_1, k_2) \neq (0, 0)$ and $0 < k_1, k_2 < N$.

  • $(k_1 f[n] + k_2 f[n + p]) \bmod M$ does not converge to zero as $n \to \infty$.
  • A non-negative integer $s = s(p) \neq p$ exists such that $(k_1 f[s] + k_2 f[s + p]) \bmod M, = \frac{M}{2}$.
  • A non-negative integer $t = t(p) < p$ exists such that $k_2 f[t] \bmod M = \frac{M}{2}$.

In the above, f is the impulse response of the forward filter $F(z)$ in the DDSM as shown in Fig. 3.3.

A corollary to this theorem is that the conditions of part (a) of the theorem are *sufficient* for the quantizer error e_q to possess the following time average properties in probability:

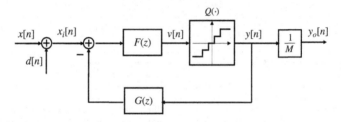

Fig. 3.3 Generic block diagram of an SQ-DDSM with LSB dithering at the input

$$\lim_{L \to \infty} S_L(e_q[n]) = M_e = \frac{1}{2}, \tag{3.1}$$

$$\lim_{L \to \infty} S_L\big((e_q[n] - M_e)x_i[n - p]\big) = 0, \tag{3.2}$$

and the conditions of part (b) of the theorem are sufficient for $e_q[n]$ to possess the following time average auto-covariance, in probability

$$\lim_{L \to \infty} S_L\big((e_q[n] - M_e)(e_q[n - p] - M_e)\big) = \sigma_{ee}^2 \delta[p],$$

where, for a given x, $S_L(x) \equiv \frac{1}{L}\sum_{n=0}^{L-1} x$ and $\sigma_{ee}^2 = \frac{M^2 - 1}{12}$.

To summarize, these conditions are sufficient to ensure that the quantization noise is uniform, white, and uncorrelated with the input in a time-averaged sense, resulting in a spectrum that is free of spurious tones.

Based on their analysis [61], Pamarti et al. have identified popular classes of SQ-DDSMs with $F(z) = z^{-l}(1 - z^{-1})^{-l}$ and $G(z) = (1 - z^{-1})^l - z^l$ which satisfy the specified conditions. The signal and noise transfer functions of these modulators are defined by

$$\text{STF}(z) = z^{-l},$$

and

$$\text{NTF}(z) = (1 - z^{-1})^{-l},$$

where l is the order of the modulator.

Their analysis [61] shows that with $l = 1$ (the first order DDSM), the modulator fails to satisfy the conditions of section (b) of the theorem. However, it satisfies the first part of section (a) of the theorem. Consequently, the quantizer error e_q converges to a uniformly distributed random variable that is asymptotically independent of x_i, but $(e_q[n], e_q[n - p])$ does not converge in distribution to the two independent random variables (U, V). For $l > 2$, the impulse response f satisfies the conditions if $M = 2^{n_0}$; therefore, the quantization noise is uniform, white, and uncorrelated with the input in a time-average sense. In addition, $F(z)$ satisfies the impulse response conditions for orders greater than two when M is a power of two.

3.2.3 LSB Dithering in the MASH DDSM

In the conventional MASH DDSM, only the quantization noise of the last stage contributes to the output spectrum. Therefore, in this case, the statistical properties of the quantization noise of the last stage are of the greatest importance.

As mentioned earlier, the addition of an LSB dither signal raises the noise floor. One method to overcome this effect, other than by choosing a large wordlength n_0, is to shape the input dither signal by passing it through a high pass filter in order to

Fig. 3.4 Block diagram of a
DDSM with shaped LSB
dithering

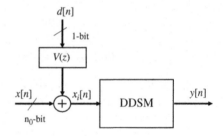

attenuate the low frequency portion of its spectrum. The shaped dither can then be
added to the input of the modulator for randomization purposes.

Figure 3.4 shows the block diagram of a DDSM with a shaped LSB dither signal
added to the input [4, 9, 37, 38, 59, 61–63].

A one-bit dither signal d is generated and is passed through a linear filter with
transfer function

$$V(z) = (1 - z^{-1})^R,$$

where R is the order of the high pass filter. The output of the filter is added to the
first few LSBs of the input x. $R = 0$ corresponds to the case of nonshaped LSB
dither.

Pamarti and Galton [9] have developed sufficient conditions so that the shaped
LSB dither signal makes the quantization noise of the last stage in the MASH
DDSM asymptotically white and independent of the input.

Theorem 1 of their work[3] specifies necessary and sufficient conditions on the
impulse responses of the forward filters $F_i(z)$ of the stages in the MASH DDSM
and the dither filter $V(z)$ to ensure that the error signals e_{qi} of the stages possess the
following ensemble statistics:

- The probability mass function (pmf)[4] of $e_{qi}[n]$ converges to that of a uniform
  random variable as $n \to \infty$.
- The conditional pmf of $e_{qi}[n]$ given $x_i[n - p]$ converges to that of a uniform
  random variable as $n \to \infty$ for every finite positive p, where $x_i[n] = x[n] +
  d[n]*v[n]$.
- The joint pmf of $e_{qi}[n]$ and $e_{qi}[n - p]$ converges to that of a pair of independent
  random variables as $n \to \infty$ for every finite $p \neq 0$.
- The joint pmf of $e_{qi}[n]$ and $e_{qj}[n - p]$ converges to that of a pair of independent
  random variables as $n \to \infty$ for every finite $i \neq j$ and for every finite integer p.

---

[3] The interested reader may consult with [9]. Here we summarize the key results.

[4] The pmf is a function that gives the probability that a discrete random variable is exactly equal to
some value. It differs from a probability density function (PDF) in that the values of a PDF, defined
only for continuous random variables, are not probabilities as such. Instead, the integral of a PDF
over a range of possible values (a,b] gives the probability of the random variable falling within that
range.

A corollary to this theorem is that the conditions of Theorem 1 are sufficient to make the following true in probability:

$$\lim_{L \to \infty} S_L(e_{qi}[n]) = M_e = \frac{1}{2} \tag{3.3}$$

$$\lim_{L \to \infty} S_L((e_{qi}[n] - M_e)x[n - p]) = 0 \tag{3.4}$$

$$\lim_{L \to \infty} S_L((e_{qi}[n] - M_e)d[n - p]) = 0 \tag{3.5}$$

$$\lim_{L \to \infty} S_L((e_{qi}[n] - M_e)(e_{qi}[n - p] - M_e)) = \sigma_{ee}^2 \delta[p] \tag{3.6}$$

$$\lim_{L \to \infty} S_L((e_{qi}[n] - M_e)(e_{qj}[n - p] - M_e)) = 0, \tag{3.7}$$

where $S_L(x) \equiv \frac{1}{L} \sum_{n=0}^{L-1} x$ and $\sigma_{ee}^2 = \frac{M^2-1}{12}$. When the properties above hold, the PSD of the output quantization noise of the MASH DDSM is free of spurious tones.

The order of the filter R cannot go beyond $l - 2$ for a given modulator order l if the above conditions are to be true (see Table 3.1). In the case of the MASH 1-1-1 DDSM, for example, the maximum acceptable value of R is 1.

Here we provide simulation results in order to gain a better understanding of the table. Figure 3.5 shows simulated output power spectral density plots of a MASH 1-1-1 DDSM with $M = 2^{n_0}$ with $n_0 = 9$, input $X = 1$, and shaped-LSB dither applied to the input. In (i), the dither is zero. In (ii), LSB dither is added directly to the input without shaping. In (iii), dither is applied to a first order filter ($R = 1$) and then added to the input, resulting in a first order shaped noise floor.

When no dither is applied, the output signal is periodic with a period of 2^{10}. This means that the quantization noise is distributed over 2^{10} tones between zero and f_s. When dither is applied directly to the modulator's input, it randomizes the quantization noise but it also raises the noise floor, as we see in (ii). As suggested, shaping the dither helps to reduce the noise floor, thereby improving the dynamic range of the system, in particular in this case where n_0 is small.

The plots in Fig. 3.5 were generated using an FFT length of $N_f = 2^{17}$ and applying a Hanning window of the same length to each signal. The power spectral density of each signal was estimated using the *periodogram* method described in the previous chapter.

Table 3.1 Valid order R of the dither filter in the shaped LSB dithering technique for randomizing the quantization noise in a class of MASH DDSMs [9]. l is the order of the MASH DDSM

Type of MASH	$|STF(z)| = 1, NTF(z)$	Filter order R for no spurs
1-1	$(1 - z^{-1})^2$	0
1-1-1, 1-2, 2-1	$(1 - z^{-1})^3$	1
$m_1 - m_2 - \cdots - m_k$	$(1 - z^{-1})^l$	$l - 2$
(where $m_1 + m_2 + \cdots + m_k = l > 3$)		

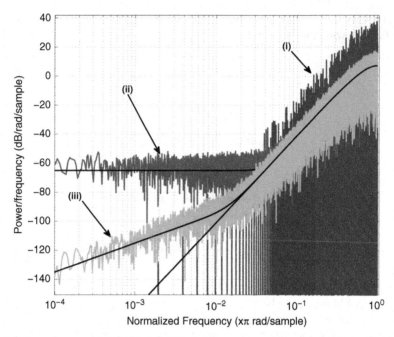

Fig. 3.5 Output spectra of a MASH 1-1-1 DDSM for three cases with constant input $X = 1$ and $M = 2^9$: (*i*) MASH without dithering, (*ii*) MASH with LSB dither added to the input and (*iii*) MASH with first-order shaped LSB dither added to the input. The *solid curves* represent shaped white quantization noise, including the dither contributions where applied

In Fig. 3.6, we show that a second order filter ($R = 2$) results in a spectrum with spurious tones, confirming the results in Table 3.1 that the maximum value of R to guarantee no spurs is unity in the case of a third order MASH DDSM.

3.2.4 Other Schemes of Dithering

Shaped LSB dithering at the input is not the only technique to randomize the quantization noise in a DDSM. Another option is to apply a dither signal to the system by adding it immediately before the quantizer [64], as illustrated in Fig. 3.7. In this way, the dither signal and the quantizer error are filtered by the same noise transfer function of the modulator. In this case, the dither spectrum does not add a noise floor to the output spectrum. However, it increases the total quantization noise, depending on the amplitude of the dither signal [64].

To randomize the quantizer error effectively, the dither signal injected directly before the quantizer must have a significant amplitude [65]. If the quantizer's step-size is denoted by M, the dither amplitude has to be close to the step-size M ($\frac{M}{2}$, M, or $2M$). This amount of dither is much greater than the LSB dither at the input. However, the noise transfer function of the modulator shapes the dither spectrum

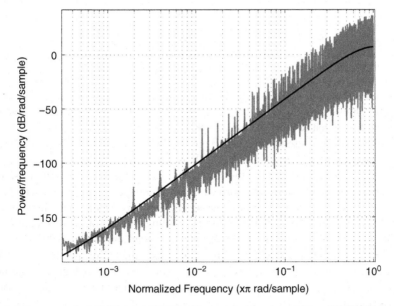

Fig. 3.6 Spectrum of a MASH 1-1-1 DDSM with second order shaped ($R = 2$) LSB dither at the input for $X = 1$, $n_0 = 9$, $M = 2^9$. In this case, the shaped dither fails to perform proper randomization, resulting in a tonal spectrum at the output. The *solid curve* represents the shaped white quantization noise including the dither contribution, assuming that the white noise approximation is valid. Note that many tones lie significantly above the approximate curve

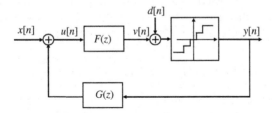

Fig. 3.7 Block diagram of a single quantizer DSM with in-loop dithering

in the same manner as it shapes the quantizer error. Nonetheless, the total shaped quantization noise at the output is increased by 3 dB if the dither amplitude is M, as estimated below.

We assume a uniform distribution for the normalized quantizer error in the range $-\frac{1}{2} \leq q \leq \frac{1}{2}$ and add uncorrelated uniformly distributed normalized digital dither in the range $-\frac{1}{2} \leq d \leq \frac{1}{2}$. The total quantization error has a triangular distribution in the range $-1 \leq q+d \leq 1$. We can calculate the variance of the total quantization error $e_t = q + d$ by writing[5]:

----

[5] q and d are discrete random variables; however, for simplicity we have assumed continuous random variables.

$$\sigma^2 = \int_{-1}^{0} (e_t + 1)e_t^2 \, de_t + \int_{0}^{1} (1 - e_t)e_t^2 \, de_t = \frac{1}{6}.$$

This value is twice the variance of the quantizer error alone; therefore, we expect that the magnitude of the power spectrum should be doubled (increased by 3 dB).

Careful simulation of each digital modulator configuration (different architecture, different order, multi-bit or single-bit quantizer) is required to find an optimum dither amplitude that renders the output spectrum free of spurious tones.

Examples of such simulations [58],[6] are explained briefly here. In [58], a third order digital error feedback modulator such as that shown in Fig. 3.8 with $H(z) = 1 - (1 - z^{-1})^3$ was considered as the case study. The authors applied a dither signal before the quantizer to study the purity of the output spectrum. The amplitude of the dither signal was chosen to be equal to the step-size of the multi-level quantizer M. Their study shows that this dither amplitude is the optimal selection for this modulator in the sense of spectral purity. A lower amplitude results in much worse tonal behavior, while a higher amplitude raises the entire spectrum envelope without significantly improving the power distribution. Also, a higher amplitude dither signal unfavorably increases the number of quantizer levels required to prevent overloading.

To illustrate this,[7] we have simulated a third order EFM (see Fig. 3.8) with $M = 2^9$ and $X = 10$ with different dither amplitudes M, $\frac{M}{2}$, $2M$, and with $H(z) = 1 - (1 - z^{-1})^3$.

The dither signals were generated using the *rand* function in MATLAB which returns a uniformly distributed random number in [0, 1]. To ensure that the dither has integer values in the range $[\frac{-M}{2}, \frac{M}{2}]$ when the dither's peak-to-peak amplitude is the same as the step-size M, we perform the following operation:

$$d_i = round(M \times rand(1, sim\_time)) - \frac{M}{2},$$

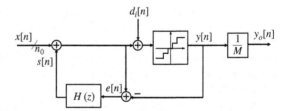

Fig. 3.8 Block diagram of a third order EFM with in-loop dithering

---

[6] The reader can refer to [4, Chap. 3, 66] for a more comprehensive study of in-loop dithering in DSMs.

[7] The detail of the quantizer implementation is not given in [58]. Here, we use a multi-level mid-tread quantizer of the type shown in Fig. 2.7b to illustrate the point.

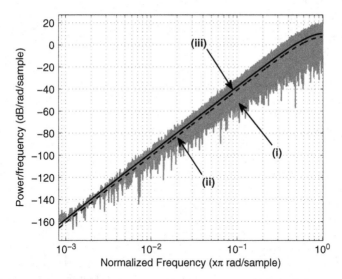

Fig. 3.9 Spectrum of a third order EFM, plot (*i*), with a uniformly distributed digital dither signal with amplitude M added to the quantizer input for $X = 10$, $n_0 = 9$, $M = 2^9$. The *solid* (*ii*) and *dashed* (*iii*) *curves* represent the shaped white quantization noise with different distributions, including the dither contribution, assuming that the white noise approximation is valid

where the function *round* rounds to the nearest integer and the *rand* function generates a random signal with a uniform distribution in [0, 1]. Figure 3.9 shows the output spectral plots of the simulated modulator with 2^{18} output samples taken for the spectral calculations. A Hanning window of the same length was used in the periodogram calculations.

There are three plots in the figure. Plot (i) shows the power spectral density of the simulated modulator. Plot (ii) (dashed curve) shows the predicted power spectral density of the output assuming ideal white quantization noise having a uniform distribution in the range $[-\frac{1}{2}, \frac{1}{2}]$. The following formula was used in order to generate this curve:

$$S(\omega_k) = \frac{1}{12} \left\{ 2 \sin \left(\frac{\omega_k}{2} \right) \right\}^6, \quad \omega_k = 0, \frac{2\pi}{N_f}, \frac{4\pi}{N_f}, \cdots, \pi. \qquad (3.8)$$

The solid curve in plot (iii) represents shaped white noise with a triangular distribution in the range $[-1\ 1]$. Analytically,

$$S(\omega_k) = \frac{1}{6} \left\{ 2 \sin \left(\frac{\omega_k}{2} \right) \right\}^6, \quad \omega_k = 0, \frac{2\pi}{N_f}, \frac{4\pi}{N_f}, \cdots, \pi. \qquad (3.9)$$

Note that the shaped quantization noise in case (iii) is doubled (increased by 3 dB) compared to case (ii).

The output occupies 13 quantizer levels ($-6M$ to $6M$) for this input and dither amplitude, as shown in the plot of the distribution of the output sequence in Fig. 3.10a. By contrast, the same modulator with first-order-shaped LSB dither at the input needs only 7 ($-3M$ to $3M$) levels, as shown in Fig. 3.10b.

Figures 3.11 and 3.12 show spectral results of the same modulator with dither amplitudes of (a) $\frac{M}{2}$ and (b) $2M$, respectively.

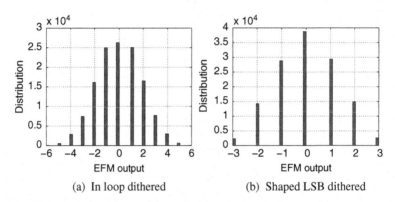

(a) In loop dithered   (b) Shaped LSB dithered

Fig. 3.10 Distribution of the output sequence of a third order EFM with (**a**) step-size in-loop dithering and (**b**) first order shaped LSB dithering at the input

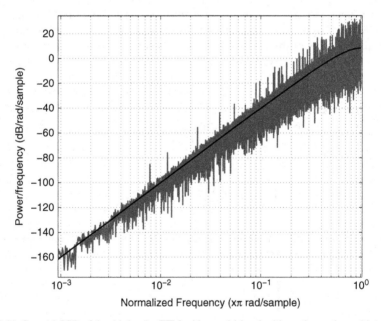

Fig. 3.11 Case (a) PSD of the third order EFM with a multi-level mid-tread quantizer with in-loop dithering amplitude of $\frac{M}{2}$. The *solid curve* represents shaped white quantization noise including the dither contribution

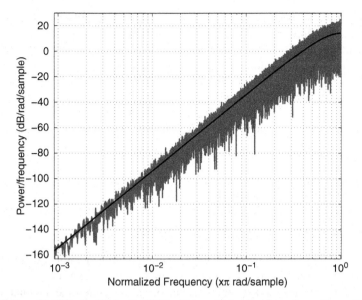

Fig. 3.12 Case (b) PSD of a third order EFM with a multi-level mid-tread quantizer with in-loop dithering amplitude of $2M$. The *solid curve* represents the shaped white quantization noise including the dither contribution

In case (a), shown in Fig. 3.11, there are many spurious tones deviating from the solid curve, suggesting that the dither amplitude of $\frac{M}{2}$ is not large enough to produce a spectrum that is free of spurious tones.

In case (b), shown in Fig. 3.12, the dither amplitude is $2M$, resulting in a larger total quantization error; consequently, the entire spectrum is shifted upwards. Also, the simulation shows that, with this dither amplitude and input, the output needs 21 ($-10M$ to $10M$) levels to prevent overloading.

In general, higher order modulators need a greater number of quantizer levels in order to work over a larger range of inputs without being overloaded. Care must be taken, in the case of in-loop dithering in particular, to provide a sufficiently large number of levels to avoid overloading the quantizer [58].

To summarize, additive noise-shaped LSB dither at the input has been shown theoretically to randomize the quantization noise in the case of MASH and SQ-DDSMs. There is a limit to the order of the filter that can be used to shape the dither signal. For an lth order MASH DDSM, the maximum order R of the shaping filter to guarantee a spur free spectrum is $(l - 2)$ [9].

As an alternative to the noise-shaped LSB dither technique, one can shape the dither spectrum with the noise transfer function of the system by adding the dither signal directly to the input of the quantizer. This dithering technique is suboptimal because it has no effect on the state variables unless it results in a change in the output bit stream. On the other hand, a small disturbance at the input, such as the described LSB dithering (continuously disturbing) method, affects all of the state variables [10]. Extensive simulation is required in order to determine an optimal dither amplitude for a given DDSM configuration. In-loop dithering increases the total spectral envelope by 3 dB if the amplitude of the dither is equal to the step-size of the quantizer.

Both of the techniques described above have been used in practice to randomize DDSMs. Care must be taken to minimize their disadvantages for a given architecture, wordlength, and modulator order.

In the remainder of this chapter, we will study deterministic techniques which randomize the quantization noise by maximizing cycle lengths without adding noise.

3.3 Deterministic Techniques

3.3.1 Setting Predefined Initial Conditions

Analog and digital Delta-Sigma modulators are prone to generating cycles [4] that result in the presence of periodic components in the output spectrum.

It has been proven [67] that continuous-amplitude discrete-time modulators can exhibit periodic behavior when rational DC inputs are applied and consequently spurious tones appear in the output power spectrum. Gray showed mathematically [67] that the output spectrum of a first order ADSM with a DC input consists of discrete spurs whose locations and amplitudes depend on the value of the input. This analysis was extended to higher order MASH topologies in [68, 69]. The results in [68, 69] demonstrate mathematically that, in a higher order[8] MASH modulator, the additive white noise assumption is asymptotically correct for the quantizer error in the last stage when *irrational* DC inputs are employed. Therefore, the higher order analog[9] MASH modulator does not have periodic behavior when an *irrational* DC input is applied. Gray et al. did not consider in their analysis effects such as gain mismatch of the stages, timing jitter, input noise, etc. Similar results have been obtained in other theoretical work on higher-order single-stage multi-bit modulators with *irrational* DC inputs [70, 71].

These mathematical analyses of analog modulators were performed with the assumption of zero initial conditions. For the case of irrational DC inputs, the asymptotic behavior of the quantizer error is not affected by the initial conditions [5, 67–69]. However, when the input has a rational DC value,[10] the initial conditions may be selected so as to randomize the quantizer error. Kozak and Kale [5] provide an exact analysis of higher order MASH DS modulators with rational DC inputs and non-zero initial conditions. Their approach shows that an irrational initial condition applied to the first integrator guarantees a tone-free output spectrum for third and higher order MASH modulators driven by rational DC inputs. Similar

---

[8] Order ≥ 3.

[9] Continuous-amplitude.

[10] In the case of DDSMs, the input is rational by definition.

analysis was performed on a class of single-quantizer modulators for which they have obtained identical results [5].

Borkowski et al. [11] have extended the results of [5] considering DDSMs from a different perspective, namely trajectories over a finite set of states. The DDSM is a finite state machine (FSM) because it is implemented using finite precision arithmetic units and a finite amount of hardware; therefore, the number of available states is finite. Without dithering, the DDSM is a deterministic FSM with a unique rule for transitioning from each state to the next. If the input is constant, the most complex behavior the DDSM can exhibit is a trajectory that visits each state once before repeating. In fact, unlike analog modulators, the output will always be constant or periodic. Therefore, the DDSM always produces a constant or a periodic output signal (a cycle) when the input is constant. In particular, the period of the signal depends on the input, the initial conditions, and the architecture of the DDSM.

Two common modulator architectures were examined in [11] to determine the effect of the initial conditions on the lengths of the cycles: (a) MASH, shown in Fig. 3.13, constructed using the EFM1 block shown in Fig. 3.14, and (b) the higher order multi-bit EFM shown in Fig. 3.15. The signal and noise transfer functions for both modulators are the same:

$$STF = 1,$$
$$NTF = (1 - z^{-1})^l, \quad (3.10)$$

where l is the order of the modulator.

The filter $H(z)$ in Fig. 3.15 (shown in detail in Fig. 3.16) can be described by:

$$H(z) = 1 - NTF, \quad (3.11)$$

where $NTF = (1 - z^{-1})^l$.

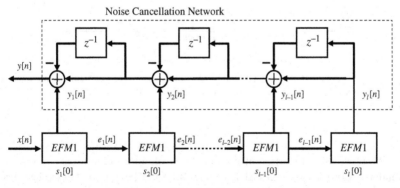

Fig. 3.13 Block diagram of a digital MASH DSM. The $EFM1$ blocks have the structure shown in Fig. 3.14. $s_i[0]$ is the initial value of the state s_i

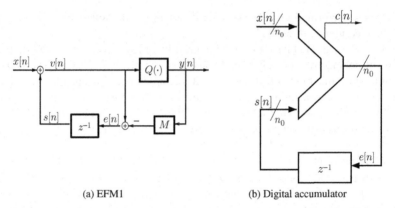

(a) EFM1   (b) Digital accumulator

Fig. 3.14 (**a**) The EFM1 block used for constructing the MASH DDSM shown in Fig. 3.13. (**b**) The digital implementation of the EFM1 block. The block $Q(\cdot)$ is a 1-bit quantizer

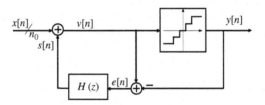

Fig. 3.15 Block diagram of a multi-bit digital EFM

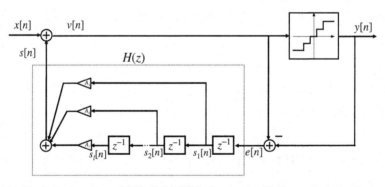

Fig. 3.16 Block diagram of a multi-bit digital EFM, showing details of a particular implementation of the filter block $H(z)$ [11]. In the case of a third order modulator, $l = 3$, $A_1 = 3$, $A_2 = -3$ and $A_3 = 1$

The authors of [11] performed extensive simulations on various orders of MASH and EFM DDSMs with different combinations of initial conditions, inputs, and accumulator word-lengths, in order to determine the lengths of the cycles. Their simulation results show that, regardless of the value of the constant input, if a certain set of initial conditions is applied, one can guarantee cycle lengths that are greater

Table 3.2 Recommended initial conditions [11] to maximize minimum cycle lengths in MASH and EFM DDSMs

Modulator order (l)	MASH	EFM
2	$s_1[0]$ odd	$s_1[0] + s_2[0]$ odd
3	$s_1[0]$ odd	$s_1[0] + s_3[0]$ odd
4	$s_1[0]$ odd	$s_1[0] + s_2[0] + s_3[0] + s_4[0]$ odd
5	$s_1[0]$ odd	$s_1[0] + s_5[0]$ odd

than a well-defined lower bound. The lower bound is a function of the accumulator word-length alone, giving the possibility to engineer the cycle length using hardware.

Table 3.2 summarizes a set of initial conditions that are recommended in [11] to maximize the cycle lengths. Table 3.3 describes the guaranteed minimum and maximum cycle lengths that can be obtained when the initial conditions in Table 3.2 are used.

The interpretation of the tables is as follows: lth order MASH and EFM topologies are considered with l in the range 2–5. First order EFM modulators (denoted EFM1 in Fig. 3.14a) are used in each stage of the MASH modulator. $s_i[0]$ denotes the initial value of the register in stage i of the MASH and the initial value of the ith register in the higher order EFM, as shown in Figs. 3.13 and 3.16. The second column in Table 3.2 recommends specific values to which the state of the first stage of the MASH should be set in order to guarantee the reported minimum cycle lengths in Table 3.3. In the case of the EFM topology, the sums of the initial conditions of various combinations of the internal registers ($s_i[0]$s) are reported in the third column in Table 3.2. For example, in the case of a second order EFM ($l = 2$), if the sum of $s_1[0]$ and $s_2[0]$ is equal to an odd value, a minimum cycle length of 2^{n_0-1} is guaranteed, regardless of the value of the constant input. By setting the initial conditions in this way, one can obtain guaranteed long cycles for all constant inputs, provided that the initial conditions are chosen from Table 3.2. In particular, by choosing a sufficiently large value of the accumulator word length n_0, one can make sure that cycles shorter than a given value are not generated. Note that cycle lengths are the same for the same order of modulator for both topologies, provided that the recommended initial conditions are set as specified. In the case of a MASH DDSM with order 3 and $M = 2^{n_0} = 2^9$, the minimum cycle length is 2^{10}, provided that an odd initial value is set in the register of the first stage (e.g. $s_1[0] = 1$).

Table 3.3 Cycle lengths in conventional MASH and EFM modulators with the initial conditions of Table 3.2 [11], as a function of the word length n_0

Modulator order l	Guaranteed minimum cycle length	Maximum cycle length
2	$2^{(n_0-1)}$	$2^{(n_0+1)}$
3	$2^{(n_0+1)}$	$2^{(n_0+1)}$
4	$2^{(n_0+1)}$	$2^{(n_0+2)}$
5	$2^{(n_0+2)}$	$2^{(n_0+2)}$

3.3.2 Example

Here we provide an example which shows the improvement one can obtain by scaling the modulator word length. Figure 3.17 shows simulations of a third order MASH 1-1-1 modulator for two cases: one with $n_0 = 9$ and the other with $n_0 = 18$. In both cases, the initial condition of the first stage is set to an odd value, as specified in Table 3.2. The input to the modulator is normalized[11] to 0.5 in both cases. According to the tables, the cycle lengths of the modulators should be 2^{10} and 2^{19}, respectively. The quantization noise power should be spread over a much larger number of tones in the latter case.

Figure 3.17 shows clearly that the spectrum of the output of the 18-bit DDSM, which has a cycle length of 2^{19}, is much smoother than that of the 9-bit modulator whose quantization noise is distributed over 2^{10} tones. The periodogram method, with a Hanning window of the same length as the output signals with 2^{19} output samples, was used in the spectral estimations.

The results concerning cycle lengths in [11] are empirical in nature. Since there is a large number of possible combinations of modulator orders, input values, initial

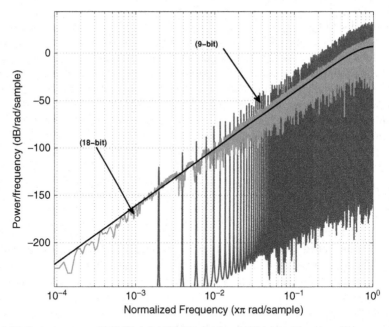

Fig. 3.17 Output spectra of MASH-1-1-1 DDSMs for 9 and 18 bit word-lengths with a normalized input of 0.5 (for the former $X = 256$ and for the latter $X = 2^{17}$). The *solid curve* represents the idealized shaped white quantization noise. An odd initial condition ($s_1[0] = 1$) is set in the first stage for both configurations

---

[11] $X = 2^8$ in the 9-bit case and $X = 2^{17}$ in the 18-bit case.

conditions, and accumulator word lengths, it will be useful to prove these results mathematically. We addressed this problem [13] for the case of MASH modulators with order in the range 1–3; the analysis is presented in Sections 3.4.1 and 3.4.2. Fitzgibbon and Kennedy have extended the proof for the MASH DDSM with orders 4 and 5 [72]. In addition to the analysis for the MASH DDSM, they also report the proof for higher order (2–5) EFMs [73].

In Section 3.4.3, we derive exact mathematical expressions for the cycle lengths as a function of the (constant) input and initial conditions for MASH DDSMs with orders 1,2 and 3.

3.4 Mathematical Analysis

In this section, we provide a mathematical analysis [13] of the MASH DDSM confirming and extending some[12] of the empirical results published in [11]. We derive nonlinear equations governing the system and then derive equations for the binary quantizer error in the last stage of a first, second and third order modulator. We perform the analysis in a step-by-step manner, starting with a first order modulator, then a MASH 1-1 modulator, and finally a MASH 1-1-1 DDSM. In each case, we use expressions for the quantizer error to calculate the period. In the analysis, we assume that an odd initial condition has been set in the first stage, as suggested in [11].

3.4.1 First Order Modulator

In this subsection, we derive a closed form expression for the error of the binary quantizer in a first order digital EFM and we calculate the corresponding cycle length (period). First order EFMs are usually used as the building blocks of MASH DDSMs due to their ease of implementation. The first order EFM can be implemented with a simple digital accumulator. The digital implementation of an EFM is shown in Fig. 3.18a and the corresponding mathematical model is shown in the block diagram in Fig. 3.18b.

The quantizer is implemented by the overflow operation in the accumulator. The carry-out bit $c[n]$ corresponds to $y[n]$. All the signals in Fig. 3.18 are integer-valued.

The input to the modulator is the signal x which has n_0-bit resolution. The 1-bit output of the modulator, weighted by $M = 2^{n_0}$, is subtracted from the quantizer input, resulting in an error signal e. The register labeled z^{-1} delays e by one clock cycle. The delayed signal s is added to the input x.

---

[12] The results in [11] are for MASH and EFM topologies with order in the range 1–5; the analysis presented here covers only the MASH DDSM with order in the range 1–3. The interested reader may consult [72] for MASH DDSM with orders 4 and 5 and [73] for higher order EFMs.

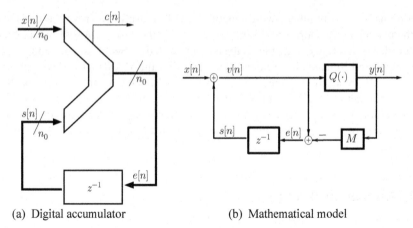

(a) Digital accumulator   (b) Mathematical model

Fig. 3.18 (**a**) The digital implementation of the EFM1 block used for constructing the MASH DDSM shown in Fig. 3.13. (**b**) The mathematical model of the EFM1 block. The block $Q(\cdot)$ is a 1-bit quantizer with the transfer function shown in Fig. 3.19

The input summing node is implemented by the adder shown in Fig. 3.18a. The quantizer block performs the following operation and its transfer characteristic is shown in Fig. 3.19:

$$c = y = Q(v) = \begin{cases} 0, & v < M \\ 1, & v \geq M. \end{cases} \tag{3.12}$$

From the time-domain perspective, the average of the output signal y, consisting of 0s and 1s, is equal to the average value of the input divided by M. For example if $n_0 = 3$, the input to the modulator is $x = 4$ and the initial condition is set to zero, a cycle $0, 1, 0, 1, 0, 1, \cdots$ of period 2 is generated, resulting in an average output of $0.5\ (= \frac{x}{M})$.

From the frequency-domain perspective, the modulator ideally implements first-order noise shaping, pushing the quantization error towards higher frequencies. In the Z-domain, the input signal passes through the system without being filtered. The signal transfer function (STF) and the noise transfer function (NTF) of the modulator are as follows:

Fig. 3.19 Input–output relationship of the quantizer shown in Fig. 3.18b

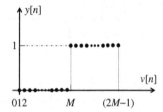

$$STF(z) = 1 \tag{3.13}$$

$$NTF(z) = (1 - z^{-1}). \tag{3.14}$$

Note that the zero of the NTF is located at $z = 1$, rejecting the DC term in the quantization noise.

For the valid range of x and s

$$x[n], \ s[n] \in \{0, 1, 2, \ldots, (2^{n_0} - 1)\}, \tag{3.15}$$

v is in the following range:

$$v[n] \in \{0, 1, 2, \ldots, (2^{n_0+1} - 2)\}. \tag{3.16}$$

The nonlinear difference equations governing the structure shown in Fig. 3.18b can be summarized as:

$$v[n] = x[n] + s[n] \ \forall \ n \geq 0, \tag{3.17}$$

$$y[n] = Q(v[n]) = \begin{cases} 0, & v[n] < M \\ 1, & v[n] \geq M, \end{cases} \tag{3.18}$$

$$s[n] = \begin{cases} e[n-1] \ \forall \ n \geq 1 \\ s[0], & n = 0, \end{cases} \tag{3.19}$$

In the following, we consider the error e defined as:

$$e[n] = v[n] - My[n]. \tag{3.20}$$

From Fig. 2.10b, the reader can confirm that $e = -Me_q$, where e_q is the quantization noise of the 1-bit quantizer.

With this definition of $e[n]$, and considering (3.16) and (3.20), its values are in the range of:

$$e[n] \in \{0, 1, 2, \ldots, (2^{n_0} - 2)\}. \tag{3.21}$$

In this case, the mean of the scaled quantization noise is non-zero, resulting in a DC term in its spectrum. Since the NTF has a zero at 1, this DC term is completely removed. However, in modulators which do not remove the quantization noise at DC completely, mid-rise or mid-tread quantizers should be used [12] since these quantizers have zero mean quantization noise.

In (3.19), $s[0]$ is the initial condition of the register in Fig. 3.18b, and we assume that $s[0]$ is an odd integer unless otherwise stated. Using the above equations, we will find an expression for $e[n]$ in terms of the input $x[n]$ and the initial condition $s[0]$ because we want to study the effect of the input and the initial condition on the periodicity of the system. To do this, we start with (3.20).

Let us consider two cases:

Case 1: If $v[n] \geq M$, using (3.18) we can write:

$$e[n] = v[n] - M, \ \forall \, n \geq 0, \tag{3.22}$$

and considering (3.16) and (3.21) we can write:

$$e[n] = v[n] \bmod M, \tag{3.23}$$

where mod is the modulo operator.
Case 2: If $v[n] < M$, using (3.18) we can write:

$$e[n] = v[n] - 0, \ \forall \, n \geq 0 \tag{3.24}$$

and considering (3.16) and (3.21), we can write:

$$e[n] = v[n] \bmod M. \tag{3.25}$$

Therefore (3.25) is valid for all $v[n]$ in its given range. Equation (3.25) is used as a starting point to find an expression for $e[n]$ in terms of $x[n]$ and $s[0]$. First, we rewrite (3.25) as:

$$e[n] = \left(x[n] + s[n] \right) \bmod M, \ \forall \, n \geq 0. \tag{3.26}$$

Rewriting the above equation and using (3.19), we have:

$$e[n] = \left(x[n] + e[n-1] \right) \bmod M, \ \forall \, n \geq 0. \tag{3.27}$$

Expanding the above equation with its indices, we obtain:

$$e[0] = \left(x[0] + s[0] \right) \bmod M, \tag{3.28}$$

$$e[1] = \left(x[1] + e[0] \right) \bmod M, \tag{3.29}$$

and, for general index n, we have:

$$e[n] = \left(x[n] + e[n-1] \right) \bmod M. \tag{3.30}$$

We substitute the value of $e[i - 1]$ in $e[i]$ from its value in the equation with index $i - 1$. Starting with index n, ending at index 0, and using the property of the modulo operator that

$$\left((a \bmod M) + b\right) \bmod M = \left(a + (b \bmod M)\right) \bmod M = (a + b) \bmod M, \quad (3.31)$$

we obtain the following expression for $e[n]$:

$$e[n] = \left(x[n] + x[n - 1] + \cdots + x[0] + s[0]\right) \bmod M \qquad (3.32)$$

$$= \left(s[0] + \sum_{k=0}^{n} x[k]\right) \bmod M. \qquad (3.33)$$

In the special case of a constant DC input X, we have:

$$e[n] = \left(s[0] + (n + 1)X\right) \bmod M, \ \forall \, n \geq 0. \qquad (3.34)$$

Note that (3.34) relates the DC input X and the initial condition $s[0]$ to the error $e[n]$. We use this to find the period (cycle length) of the signal. If the signal e has a fundamental period N_1, then

$$e[n] = e[n+N_1] = \left(s[0]+(n+N_1+1)X\right) \bmod M = \left(s[0]+(n+1)X+N_1X\right) \bmod M,$$
$$(3.35)$$

and therefore the term $N_1 X$ should be an integer multiple of M so that $e[n] = e[n + N_1]$ is correct. Equivalently,

$$N_1 X \bmod M = 0. \qquad (3.36)$$

Because M is even, the maximum cycle length ($N_1 = M = 2^{n_0}$) is only achieved when the input X is an odd number.

3.4.1.1 Spectral Analysis

We have shown mathematically that the maximum cycle length is $M = 2^{n_0}$ in the case of a first order EFM with a binary quantizer and an odd input, regardless of the initial condition.

The first order EFM performs first order shaping of the quantization noise. If the quantization noise is assumed white, the resulting noise spectrum is gently shaped toward high frequencies. The question here is: do we get a smoothly shaped spectrum when we apply an odd input? As we will see through simulations, the answer

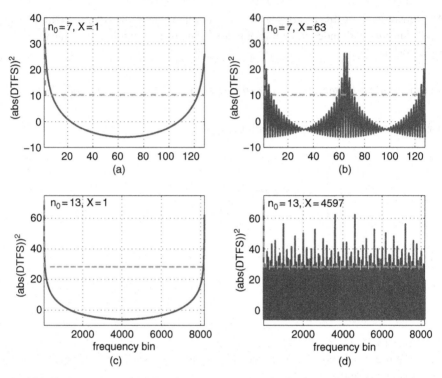

Fig. 3.20 Simulated spectra (in dB) of the error signal e of a first order EFM. (a) $n_0 = 7$ and $X = 1$, (b) $n_0 = 7$ and $X = 63$, (c) $n_0 = 13$ and $X = 1$, (d) $n_0 = 13$ and $X = 4597$. The idealized *dashed curves* are defined by (3.39)

is no. In fact, even with a maximum cycle length, the quantizer error of the first order EFM is not white due to insufficient randomization within the periods.

Here, we provide sample spectral plots to illustrate this point. Figure 3.20 shows the squared magnitude of the Discrete Time Fourier Series (DTFS) coefficients[13] of the error e in a first order EFM.

The following simulation parameters are used: (a) accumulator word length $n_0 = 7$ ($M = 2^7$) and input $X = 1$; (b) $n_0 = 7$ and $X = 63$; (c) $n_0 = 13$ and $X = 1$ and (d) $n_0 = 13$ and $X = 4597$. One full cycle of the error signal e was used in each DTFS calculation.

The DTFS of $x[n]$ is calculated in MATLAB using the *fft* command which implements the following equation:

$$\text{FFT}[n] = \sum_{k=0}^{N-1} x[n] e^{-\frac{j2\pi kn}{N}} \text{ for } 0 \leq n < N. \tag{3.37}$$

---

[13] This gives the exact power per tone in the spectrum of a periodic signal.

We can calculate the DTFS coefficients as follows:

$$\text{DTFS}[n] = \frac{1}{N}\text{FFT}[n] \text{ for } 0 \le n < N.$$ (3.38)

In all four cases, the input is odd; therefore, the cycle length achieves the maximum value M. However, the spectra deviate significantly from the idealized spectra (dashed lines). In cases (b) and (c), this deviation manifests itself as spurs.

The dashed lines in Fig. 3.20 were plotted using the following formula [12, pp. 101–103]:

$$P_{e-\text{EFM}} = \begin{cases} \dfrac{(M-1)^2}{4}, & \text{at DC} \\ \dfrac{(M^2-1)}{12M}, & \text{at nonzero frequency bins.} \end{cases}$$ (3.39)

As mentioned before, there is a large DC component in the error spectrum. This is removed by the NTF since it has a zero at $z = 1$.

Figure 3.21 shows the DTFS of the output signal y with the same parameters. When the input is 1 in both cases ($n_0 = 7, 13$), y contains one 1 and the rest of the output sequence is zero within each period. Since the DTFS of an impulse signal is a flat spectrum, we would expect to have flat spectra in these two cases. Consequently, the modulator fails to accomplish noise shaping, as illustrated in subplots (a) and (c) of Fig. 3.21.

For inputs 63 and 4596, however, the modulator generates 1s and 0s more randomly within the given period; therefore, we can see the effect of noise shaping that attenuates the tones at lower frequencies and amplifies the tones at higher frequencies. However, the modulator is still not able to randomize the sequence sufficiently to produce a smooth spectrum. Therefore, as we see from the spectral plots, they still contain high power tones.

The idealized dashed curves are defined by:

$$P_{esh-\text{EFM}} = \begin{cases} 0, & \text{at DC} \\ \dfrac{1}{M^2}\dfrac{(M^2-1)}{12M}\left(2\sin\left(\dfrac{\omega}{2}\right)\right)^2, & \text{at nonzero frequency bins.} \end{cases}$$ (3.40)

Simulations show that higher order modulators do a better job of randomization, both with deterministic and stochastic techniques, as we will illustrate in the next subsection.

3.4.2 Higher Order Modulators

In the last subsection, we derived an expression for the quantization error of a first order EFM in terms of its input and initial condition, and we found the cycle length using that equation. We concluded that the maximum cycle length $N_1 = 2^{n_0}$ is

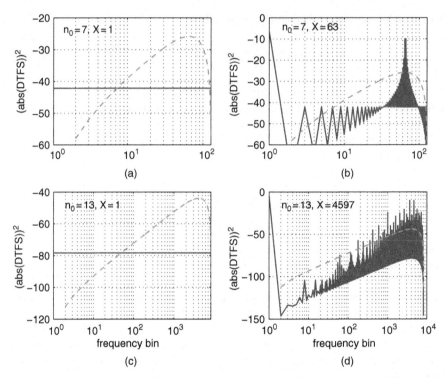

Fig. 3.21 Simulated spectra (in dB) at the output of a first order EFM. (**a**) $n_0 = 7$ and $X = 1$, (**b**) $n_0 = 7$ and $X = 63$, (**c**) $n_0 = 13$ and $X = 1$, (**d**) $n_0 = 13$ and $X = 4597$. The idealized *dashed curves* are defined by (3.40)

achieved when the input is odd, regardless of the initial condition. We observed through spectral simulations that a maximum cycle length in a first order EFM is not sufficient to obtain a smoothly shaped spectrum. In this sense, higher order modulators are closer to the ideal because they do a better job of randomization within periods.

Higher order MASH DDSMs are usually composed of the structure (the first order EFM) investigated in the previous section (the $EFM1$ blocks in Fig. 3.13).

The general structure of an lth order MASH DDSM is shown in Fig. 3.13 and is repeated in Fig. 3.22 for the reader's convenience. The structure consists of several cascaded first order EFMs and a noise cancellation network. The input to stage j is the quantizer error e_{j-1} of stage $j - 1$. The noise cancellation network is used to eliminate the quantizer errors of all stages except for the last one.

The spectrally-shaped quantizer error of the last stage appears directly in the output and therefore we can derive a closed form expression for the binary quantizer error of the last stage by using the method presented in the previous section. In order to do that, we use (3.34) and (3.33) to relate the quantizer error of each stage to its input (which is the error from the previous stage):

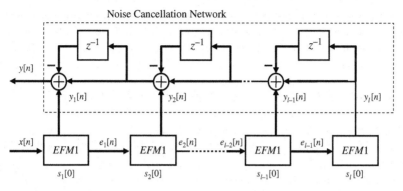

Fig. 3.22 Block diagram of a MASH DDSM. $EFM1$ blocks have the structure shown in Fig. 3.14. $s_i[0]$ is the initial value of the signal s_i

$$e_1[n] = \left(s_1[0] + (n+1)X \right) \bmod M,$$

$$e_2[n] = \left(s_2[0] + \sum_{k=0}^{n} e_1[k] \right) \bmod M,$$

$$\vdots$$

$$e_l[n] = \left(s_l[0] + \sum_{k=0}^{n} e_{l-1}[k] \right) \bmod M, \tag{3.41}$$

where $s_i[0]$ is the initial condition of stage i. We can eliminate the intermediate variables $e_i[n]$ ($i = 1, \cdots, (l-1)$) using (3.31) to arrive at an expression for $e_l[n]$:

$$e_l[n] = \left(s_l[0] + \sum_{k_{l-1}=0}^{n} s_{l-1}[0] + \sum_{k_{l-2}=0}^{k_{l-1}} s_{l-2}[0] + \cdots + \sum_{k_1=0}^{k_2} s_1[0] + (k_1+1)X \right) \bmod M. \tag{3.42}$$

Note that the error of the last stage is a function of the input X and the initial conditions of all stages. In the following, we simplify the above equation for the cases of second and third order modulators and we will use the resulting simplified equations to prove the result presented empirically in [11].

3.4.2.1 Second Order Modulator ($l = 2$)

We can determine the cycle length of the second order modulator by applying (3.42). After expanding (3.42), we obtain the simplified equation:

$$e_2[n] = \left(s_2[0] + \sum_{k=0}^{n} s_1[0] + (k+1)X \right) \bmod M, \tag{3.43}$$

that relates the error signal e_2 to the input and the initial conditions. In order to find the period N_2 of the second order system (MASH 1-1), we put the following constraint on the equation

$$e_2[n] = e_2[n + N_2] = \left(s_2[0] + \sum_{k=0}^{n+N_2} (s_1[0] + (k+1)X) \right) \bmod M \qquad (3.44)$$

$$= \left(s_2[0] + \sum_{k=0}^{n} (s_1[0] + (k+1)X) + \sum_{k=n+1}^{n+N_2} (s_1[0] + (k+1)X) \right) \bmod M.$$

$$(3.45)$$

The last term should be an integer multiple of M so that the equality holds:

$$\sum_{k=n+1}^{n+N_2} s_1[0] + (k+1)X = 0 \mod M, \forall n \geq 0. \qquad (3.46)$$

Further expansion of this expression yields the condition

$$N_2 \left(s_1[0] + \left(N_2 + (2n+3) \right) \frac{X}{2} \right) = 0 \mod M, \forall n \geq 0, \qquad (3.47)$$

where $s_1[0]$ is the initial condition on the first stage and X is the DC input. Note that the period N_2 of the MASH 1-1 error does not depend on the initial condition of the second stage.

We assume that $s_1[0]$ is odd, as suggested in [11]. Therefore, we examine the periodicity of (3.47) with an odd initial condition. We consider three cases, depending on X:

Case 1: X is an odd integer. Since the term $\left(N_2 + (2n+3) \right)$ is an odd integer,[14] it is not divisible by 2. Thus, N_2 has to be at least $2M$ so that the term $\frac{N_2}{M} \left(N_2 + (2n+3) \right) \frac{X}{2}$ is an integer. Therefore, the minimum non-zero solution for N_2 is $2M$.

When the input is even, we consider the following two cases (2 and 3):
Case 2: Let X be of the form $2k$. When k is odd, we can write the left hand side of (3.47) as:

$$N_2 \left(s_1[0] + \left(N_2 + (2n+3) \right) k \right). \qquad (3.48)$$

The expression $\left(s_1[0] + \left(N_2 + (2n+3) \right) k \right)$ is even (because k and $s_1[0]$ are odd); therefore, the minimum value of N_2 is $\frac{M}{2}$ for all $n \geq 0$.

---

[14] M is a power of two; therefore, the period cannot be odd.

Table 3.4 The cycle lengths N_2 of the quantization error in a MASH 1-1 DDSM as a function of the digital input word X and initial condition $s_1[0]$ of the first stage

$s_1[0]$	X	N_2
Odd	Odd	$2M = 2^{n_0+1}$
Odd	$4k, k$ is integer	$M = 2^{n_0}$
Odd	$2k, k$ is odd	$\frac{M}{2} = 2^{n_0-1}$

Case 3: If the input is in the form of $4k$, we can write the left hand side of (3.47) as:

$$N_2\big(s_1[0] + \big(N_2 + (2n+3)\big)2k\big). \qquad (3.49)$$

Since the term $\big(s_1[0] + \big(N_2 + (2n+3)\big)2k\big)$ is odd, the minimum solution for N_2 is M in this case. Table 3.4 summarizes the resulting fundamental periods of the quantization error in the second order modulator.

Note that the worst case (minimum possible) cycle length is $\frac{M}{2}$ when the initial condition is odd. The maximum cycle length $2M$ is achieved when the input and the initial condition of the first stage are both odd.

3.4.2.2 Spectral Investigation

Here we discuss the effect on the spectral performance of the modulator of seeding the first stage with an odd integer. The cycle length of a MASH 1-1 DDSM with an odd seed value in the first stage is one of the three values $\frac{M}{2}$, M or $2M$, depending on the input value. By choosing a large value for n_0, one can obtain a large cycle length. Again, the question is: can one obtain a spurious-tone free spectrum by maximizing the cycle length in a second order modulator? As in the case of a first order modulator, we address this question through spectral simulations.

Figure 3.23 shows simulation results for two cases: (i) a MASH 1-1 DDSM with an odd seed value and without dither; and (ii) the same modulator with non-shaped LSB dithering at the input.

Recall from Table 3.1 that the maximum shaping order for a MASH 1-1 DDSM is $R = 0$, meaning that only non-shaped LSB dither can randomize the quantization error effectively.

These simulations were performed on a second order MASH DDSM with $n_0 = 17$, $s_1[0] = 51$, $s_2[0] = 0$, and the input $X = \frac{M}{2} + \frac{M}{4} + \frac{M}{8}$ in both cases. Plot (i) shows the spectrum for the case without dither and plot (ii) shows the spectrum for the case with dither. Note that the nonshaped LSB dither at the input randomizes the quantization error effectively in the second order modulator, resulting in a spectrum without spurious tones. However, it raises the noise floor to -115 dB. Note that the increase in the noise floor is not significant because we have used a large value for n_0, namely 17 in this example.

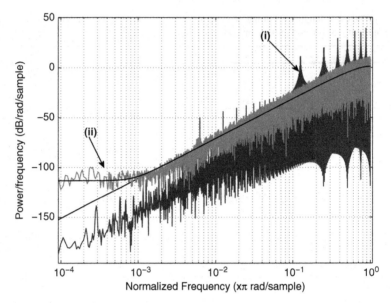

Fig. 3.23 Output spectra of a MASH 1-1 DDSM with $n_0 = 17$, $s_1[0] = 51$, $s_2[0] = 0$, and $X = \frac{M}{2} + \frac{M}{4} + \frac{M}{8}$ for two cases: (i) seeded, (ii) dithered MASH DDSMs. The *solid curves* represent the shaped white quantization noise

Despite having a cycle length of 2^{17}, the seeding technique is not as effective as the dithering method in approximating the desired ideal spectrum; the resulting spectrum contains undesirable spurious tones. This problem has been predicted theoretically in [5]. In the case of a ditherless analog modulator, even with an irrational initial condition in the first stage, the second order modulator partially decorrelates the last stage's quantizer error; however, it still fails to satisfy the assumption of the white noise model. Kozak and Kale suggest using a MASH DDSM with order higher than 2 in order to obtain spurious-tone free spectra without dither.

3.4.2.3 Third Order Modulator ($l = 3$)

Next we determine the cycle length in the case of a MASH 1-1-1 DDSM. We expand (3.42) for the third order modulator as follows:

$$
\begin{aligned}
e_3[n] &= \left(s_3[0] + \sum_{k_2=0}^{n} s_2[0] + \sum_{k_1=0}^{k_2} (s_1[0] + (k_1 + 1)X) \right) \bmod M \\
&= \left(s_3[0] + (n+1)s_2[0] + \frac{(n+1)(n+2)}{2} s_1[0] \right. \\
&\quad \left. + X \frac{(n+1)(n+2)(n+3)}{6} \right) \bmod M.
\end{aligned}
\tag{3.50}
$$

In arriving at (3.50), we have used the following identities:

$$\sum_{i=1}^{n} i = \frac{n(n+1)}{2}, \tag{3.51}$$

$$\sum_{i=1}^{n} i^2 = \frac{n(n+1)(2n+1)}{6}. \tag{3.52}$$

The error of the third stage is a nonlinear function of the input X and the initial conditions of all three stages. As we will see, the period of the signal e_3 does not depend on the initial condition of the last stage. If the fundamental period of e_3 is N_3, then the following condition holds:

$$e_3[N_3] = e_3[2N_3]. \tag{3.53}$$

Applying this condition and following the same procedure used for the second order case, we find that the sum (3.54) should be an integer multiple of M so that (3.53) holds. Thus,

$$N_3 s_2[0] + \frac{3N_3(N_3+1)}{2} s_1[0] + X N_3(N_3+1)(N_3+2) + X\left(\frac{N_3(N_3-1)(N_3+1)}{6}\right), \tag{3.54}$$

must be an integer multiple of M. Equivalently,

$$\frac{N_3 s_2[0]}{M} + \frac{3N_3(N_3+1)}{2M} s_1[0] + \frac{X N_3(N_3+1)(N_3+2)}{M}$$
$$+ \frac{X}{M}\left(\frac{N_3(N_3-1)(N_3+1)}{6}\right) = \frac{N_3 s_2[0]}{M} + a + b + c \tag{3.55}$$

must be an integer. Equation (3.55) shows that the cycle length does not depend on the initial condition of the last stage in this case.

For simplicity, we assume that $s_2[0]$ is zero and label the remaining three terms a, b and c, respectively. We need to find N_3 such that (3.55) is an integer. The solution is primarily determined by the term a. Since $s_1[0]$ is assumed to have an odd value, the minimum value for N_3 so that a is an integer is $2M$ because $s_1[0](N_3+1)$ is odd. This required minimum solution for a is sufficient to ensure that the other two terms are integers. Note that $(N_3-1)(N_3+1)$ is divisible by 3 in the last term since N_3 is a power of 2.

In the case of a non-zero initial condition in the second stage ($s_2[0] \neq 0$), and considering (3.55), this solution for N_3 is still valid. Therefore, if $s_1[0]$ is odd, the initial condition of the second stage does not affect the cycle length either.

3.4.2.4 Spectral Investigation

As in the previous case, we consider the effect of the maximized cycle length on the MASH 1-1-1 DDSM. With the same modulator word length $n_0 = 17$, the odd seeded MASH 1-1-1 DDSM generates output signals with periods of 2^{18}. Here we present MATLAB simulations in order to illustrate the effect of cycle length.

Figure 3.24 shows power spectra of the output of a third order MASH DDSM with $n_0 = 17$, $X = \frac{M}{2} + \frac{M}{4} + \frac{M}{8}$, $s_1[0] = 51$, $s_2[0] = 0$, and $s_3[0] = 0$ for two cases: (i) shaped-LSB-dithered ($R = 1$) and (ii) odd-seeded MASH DDSMs. As proven in [9], the shaped-LSB-dithered MASH modulator has a spectrum that is free of high power spurious tones. Unlike the second order modulator, the seeded MASH DDSM generates a spectrum that is spurious-tone free with a 60 dB/decade slope.

In continuation of their theoretical work on analog modulators with irrational initial conditions in the first stage, Kozak and Kale [5] have performed extensive simulations on MASH DDSMs of order 3–5 with odd initial conditions in the first stage. They concluded that the MASH DDSM with order higher than 2 and with an accumulator word length greater than 14 generates a spectrum that is free of spurious tones; this is consistent with our simulations.

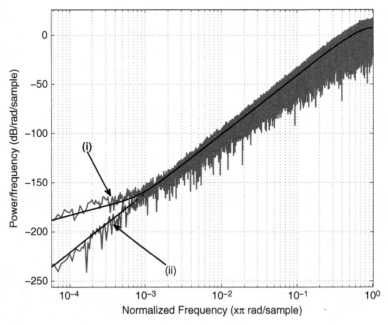

Fig. 3.24 Output spectra of a MASH 1-1-1 DDSM with $n_0 = 17$, $s_1[0] = 51$, $s_{2,3}[0] = 0$, and $X = \frac{M}{2} + \frac{M}{4} + \frac{M}{8}$ for two cases: (*i*) dithered and (*ii*) seeded MASH DDSMs. The *solid curves* represent the shaped quantization noise and dither

3.4.3 Calculation of Cycle Lengths for MASH DDSMs

Having proven the empirical results in [11] for the case of MASH DDSMs of order 2 and 3, we next calculate the cycle lengths exactly for a given order, a constant input, and a prescribed set of initial conditions. In the following section, we calculate the lengths of the cycles for MASH DDSMs of orders 1–3. Knowledge of the exact cycle lengths for undithered DDSMs with respect to the input, initial conditions, and quantizer modulus can enable a designer to predict the positions of spurs in the spectrum of the output [74].

In the remainder of this section, exact formulae for calculating cycle lengths with all possible constant inputs, initial conditions and moduli are provided for first and second order modulators. In the case of a third order modulator, the calculation becomes complicated and the cycle length is presented only for the special case when it is not divisible by 3 but it is divisible by 4.

We investigate the MASH DDSM shown in Fig. 3.25 to calculate its cycle lengths analytically in terms of the input and initial conditions. For this modulator, STF $= \frac{1}{M} z^{-(l-1)}$ and NTF $= (1 - z^{-1})^l$. We take $s_i[n]$ as the state variable and we develop equations to calculate the cycle lengths of the quantization error in the first, second and third stages, respectively. We exploit the constraints on each stage imposed by the previous stages.

3.4.3.1 Calculation of Exact Cycle Length for the First Stage

Referring to the previous analysis, error feedback is a modulo operation given by:

$$s[n] = v[n-1] \bmod M, \quad s_1[n], v[n], M \in \mathbb{N}, \tag{3.56}$$

where M is the step size of the 1-bit quantizer. In the previous analysis, we assumed that $M = 2^{n_0}$. Here, we allow M to have an arbitrary integer value.

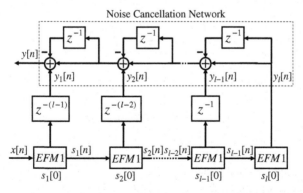

Fig. 3.25 Block diagram of the MASH DDSM analyzed in Section 3.4.3. $EFM1$ blocks have the structure shown in Fig. 3.26. $s_i[0]$ is the initial value of the signal s_i. STF $= \frac{1}{M} z^{-(l-1)}$ and NTF $= (1 - z^{-1})^l$

Fig. 3.26 Block diagram of a
first order error feedback
modulator used in Fig. 3.25

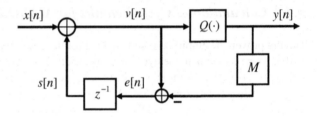

Fig. 3.26 Block diagram of a first order error feedback modulator used in Fig. 3.25

$s[n]$, $v[n]$ and M are shown in Fig. 3.26 and are integers. From Fig. 3.26, note that $v[n] = x[n] + s[n]$.

Expanding (3.56) over its indices to find $s[n]$ with respect to $x[n]$ and the initial condition $s[0]$, we obtain:

$$s[1] = (x[0] + s[0]) \mod M,$$
$$s[2] = (x[1] + s[1]) \mod M,$$

$$\vdots$$

$$s[k-1] = (x[k-2] + s[k-2]) \mod M,$$
$$s[k] = (x[k-1] + s[k-1]) \mod M. \tag{3.57}$$

To simplify this set of equations and to eliminate the intermediate indices, the value of $s[i-1]$ in each equation is substituted from the previous equation. Doing so, and using the property of the modulo operator that

$$\left(a + (b+c) \mod M\right) \mod M = (a + b + c) \mod M, \tag{3.58}$$

we obtain

$$s[n] = \left(s[0] + \sum_{k=0}^{n-1} x[k]\right) \mod M. \tag{3.59}$$

In the case of a constant input X, we have:

$$s[n] = \left(s[0] + nX\right) \mod M. \tag{3.60}$$

We start with this equation to find the cycle length L_{s1} of the first stage. If s_1 is periodic with a fundamental period L_{s1}, then $s_1[n + L_{s1}] = s_1[n]$ for all n. Without loss of generality, we require that $s_1[0] = s_1[L_{s1}]$. Thus,

$$s_1[0] = \left(s_1[0] + L_{s1}X\right) \mod M. \tag{3.61}$$

This implies that

$$L_{s1}X = 0 \bmod M. \tag{3.62}$$

The smallest value for L_{s1} that satisfies (3.62) is

$$L_{s1} = \frac{M}{\mathrm{GCD}(X, M)}, \tag{3.63}$$

where $\mathrm{GCD}(a,b)$ is the greatest common divisor of the integers a and b.

If X and M are co-prime,[15] then the maximum value of L_{s1} is M; otherwise, $L_{s1} < M$. From (3.62), we note that the cycle length of the first stage does not depend on the initial condition $s_1[0]$ but it does depend on the input X. If $X > 1$, then the minimum cycle length occurs when $X = \frac{M}{2}$, in which case $L_{s1} = 2$.

3.4.3.2 Calculation of Cycle Length for the Second Stage

We use (3.59) to find an expression for $s_2[n]$ and we then solve an appropriate equation to find the cycle length L_{s2} of this stage. Using Eqs. (3.58), (3.59) and (3.60), we write for the second stage:

$$s_2[n] = \left(s_2[0] + \sum_{k=0}^{n-1} (s_1[0] + kX) \right) \bmod M. \tag{3.64}$$

Simplifying the above equation,

$$s_2[n] = \left(s_2[0] + s_1[0]n + \frac{1}{2}Xn(n-1) \right) \bmod M, \tag{3.65}$$

and, without loss of generality, imposing the condition $s_2[L_{s2}] = s_2[0]$ to determine L_{s2}, we find that

$$\left(L_{s2}s_1[0] + \frac{1}{2}XL_{s2}(L_{s2} - 1) \right) \bmod M = 0. \tag{3.66}$$

L_{s2} must also satisfy (3.62); hence, L_{s2} must be an integer multiple of L_{s1}. Therefore, we look for the minimum value for L_{s2} as the period of the second stage; namely,

$$L_{s2} = n_{min}L_{s1}, \quad n_{min}, L_{s2}, L_{s1} \in \mathbb{N}. \tag{3.67}$$

The second term in (3.66) can be simplified in the two cases where L_{s1} is odd or even (Table 3.5). The second row of the table is derived by considering an odd value

---

[15] Two integers a and b are said to be co-prime if they have no common positive factor other than 1.

Table 3.5 The values of the second term in (3.66) for L_{s1} odd and even

L_{s1}	$\frac{1}{2}XL_{s2}(L_{s2}-1)$
Odd	$0 \bmod M$
Even	$-\frac{1}{2}XL_{s2} \bmod M$

for L_{s1}, odd $(2k+1)$ and even $(2k)$ values for n_{min}, and $(L_{s1}X) \bmod M = 0$ (see (3.62)). The last row uses (3.62) with L_{s1} even.

In the following, we find the solution for L_{s2} by considering these two cases separately:

Case 1: L_{s1} **is odd**. In this case, we have:

$$\left(L_{s2}s_1[0]\right) \bmod M = 0. \tag{3.68}$$

Therefore,

$$\left(n_{min}L_{s1}s_1[0]\right) \bmod M = 0, \tag{3.69}$$

and

$$n_{min} = \left(\frac{M}{GCD(M, L_{s1}s_1[0])}\right). \tag{3.70}$$

This gives the formula for L_{s2}:

$$L_{s2} = \left(\frac{M}{GCD(M, L_{s1}s_1[0])}\right)L_{s1} \tag{3.71}$$

$$= \left(\frac{M}{GCD(M, L_{s1}s_1[0])}\right)\left(\frac{M}{GCD(M, X)}\right).$$

Case 2: L_{s1} **is even**. Using Table 3.5, we rewrite (3.66) as:

$$\left(n_{min}L_{s1}(s_1[0] - \frac{1}{2}X)\right) \bmod M = 0, \tag{3.72}$$

$$\left(n_{min}L_{s1}(2s_1[0] - X)\right) \bmod 2M = 0, \tag{3.73}$$

$$n_{min} = \left(\frac{2M}{GCD(2M, |2s_1[0] - X|L_{s1})}\right), \tag{3.74}$$

and

$$L_{s2} = \left(\frac{2M}{GCD(2M, |2s_1[0] - X|L_{s1})}\right)L_{s1} \tag{3.75}$$

$$= \left(\frac{2M}{GCD(2M, |2s_1[0] - X|L_{s1})}\right)\left(\frac{M}{GCD(M, X)}\right).$$

From the above equations for L_{s2}, we note that the cycle length does not depend on the initial condition of the second stage. However, the initial condition of the first stage, the input X, and the modulo M of the modulator all play essential roles in determining the cycle length L_{s2}. Also, from (3.71), we find that even choosing a prime value for M does not increase the cycle length; in that case, the cycle length of the second stage is still only M, as for the first stage.

3.4.3.3 Calculation of Cycle Length for the Third Stage

We write $s_3[n]$ as

$$s_3[n] = \left(s_3[0] + \sum_{k_2=0}^{n-1} \left(s_2[0] + \sum_{k_1=0}^{k_2-1} (s_1[0] + k_1 X) \right) \right) \bmod M. \tag{3.76}$$

Simplifying the above equation, we obtain

$$s_3[n] = \left(s_3[0] + ns_2[0] + \frac{1}{2} s_1[0]n(n-1) + \frac{1}{6} Xn(n-1)(n-2) \right) \bmod M. \tag{3.77}$$

Forcing $s_{s3}[L_{s3}] = s_3[0]$, we obtain the following equation:

$$\left(L_{s3}s_2[0] + \frac{1}{2} s_1[0]L_{s3}(L_{s3}-1) + \frac{1}{6} XL_{s3}(L_{s3}-1)(L_{s3}-2) \right) \bmod M = 0. \tag{3.78}$$

We solve (3.78) in the special case when L_{s3} is not divisible by 3, L_{s2} is even, and L_{s3} is divisible by 4. If L_{s3} is not divisible by 3, $(L_{s3}-1)(L_{s3}-2)$ is always divisible by 3 for $L_{s3} > 2$ and, since $L_{s3}X \bmod M = 0$, the last term in (3.78) modulo M is equal to zero. If L_{s3} is divisible by 4, adding $-\frac{L_{s3}}{4} L_{s3}X$ to the second term and using (3.72), we obtain

$$\underbrace{s_1[0]\left(\frac{L_{s3}}{2}\right)L_{s3} - \frac{L_{s3}}{4} L_{s3}X}_{\text{mod } M, \ =0} - s_1[0]\left(\frac{L_{s3}}{2}\right) = -s_1[0]\left(\frac{L_{s3}}{2}\right) \bmod M. \tag{3.79}$$

Now we are ready to solve (3.78). We have

$$\left(L_{s3}s_2[0] - \frac{1}{2} s_1[0]L_{s3} \right) \bmod M = 0, \tag{3.80}$$

and

$$\left(n_{min} L_{s2}\left(s_2[0] - \frac{1}{2}s_1[0]\right)\right) \bmod M = 0, \ n_{min} \in \mathbb{N}. \tag{3.81}$$

Therefore,

$$n_{min} = \left(\frac{2M}{\text{GCD}(2M, |2s_2[0] - s_1[0]|L_{s2})}\right). \tag{3.82}$$

The final solution for L_{s3} is:

$$L_{s3} = \left(\frac{2M}{\text{GCD}(2M, |2s_2[0] - s_1[0]|L_{s2})}\right) L_{s2}, \tag{3.83}$$

where L_{s2} is calculated from (3.75). The results are summarized in Table 3.6.

We have verified (3.83) by simulation for the special case $M = 2^{n_0}$, with $n_0 = 3$, 4, 5, 6, 7 for all inputs and for all initial conditions. Figure 3.27 shows an example of the cycle length L_{s3} as a function of the input for $M = 2^4$ with initial conditions $s_1[0] = 8$ and $s_2[0] = 0$. Note that the worst case produces a cycle of period 4 when the input is 8, as predicted in Table 3.6.

Now that we have derived expressions to calculate the cycle length exactly in terms of M, the initial conditions, and the input X, we will validate our previous analysis of calculating the cycle length with an odd initial condition in the first stage ($s_1[0]$ is odd). The equation for calculating L_{s3} was derived based on the assumption that L_{s3} is not divisible by 3 and that it is divisible by 4 (see Table 3.6), a condition which is satisfied when $M = 2^{n_0}$ and $n_0 \geq 2$.

First, we calculate L_{s1}. We can write the input X as

$$X = K2^{n_1},$$

Table 3.6 Cycle length as a function of M, X and initial conditions $s_1[0]$, $s_2[0]$ in a third order MASH DDSM comprising first order EFMs

Stage i	Cycle length Lsi	Condition(s)
1	$L_{s1} = \left(\dfrac{M}{\text{GCD}(M,X)}\right)$	
2	$L_{s2} = \left(\dfrac{M}{\text{GCD}(M,L_{s1}s_1[0])}\right) L_{s1}$	L_{s1} is odd
2	$L_{s2} = \left(\dfrac{2M}{\text{GCD}(2M,|2s_1[0]-X|L_{s1})}\right) L_{s1}$	L_{s1} is even
3	$L_{s3} = \left(\dfrac{2M}{\text{GCD}(2M,|2s_2[0]-s_1[0]|L_{s2})}\right) L_{s2}$	

when L_{s2} is even, L_{s3} is divisible by 4 but not by 3.

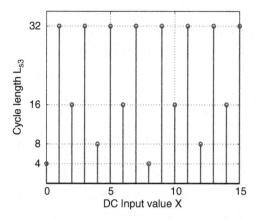

Fig. 3.27 Cycle length L_{s3} versus input in a MASH 1-1-1 DDSM with $s_1[0] = 8$, $s_{2,3}[0] = 0$ and $M = 2^4$

where K is an odd integer, and n_1 is an integer (including 0). If $n_1 = 0$, i.e. X is odd, then

$$L_{s1} = M.$$

If $n_1 \neq 0$, then

$$L_{s1} = 2^{n_0 - n_1}, \tag{3.84}$$

because $\text{GCD}(M, K2^{n_1}) = 2^{n_1}$ in this case.

In order to calculate L_{s2}, we consider the following three cases:

Case 1: $n_1 = 0$ ($X = K$ where K is odd).

In this case, $L_{s1} = M$ and, using Table 3.6, we can write L_{s2} as

$$L_{s2} = \left(\frac{2M}{\text{GCD}(2M, (|2s_1[0] - K|)M)} \right) M.$$

Note that with an odd K, $(|2s_1[0] - K|)$ is an odd integer; therefore, $\text{GCD}(2M, (|2s_1[0] - K|)M) = M$, resulting in $L_{s2} = 2M$. This confirms the first row of Table 3.4.

Case 2: $n_1 = 1$ ($X = 2K$ with K being odd).

According to (3.84), $L_{s1} = \frac{M}{2}$. In this case, we write L_{s2} as

$$L_{s2} = \left(\frac{2M}{\text{GCD}(2M, (2|s_1[0] - K|)\frac{M}{2})} \right) \frac{M}{2}.$$

Since $(|s_1[0] - K|)$ is even,[16] $\mathrm{GCD}(2M, (|s_1[0] - K|)M) = 2M$, resulting in $L_{s2} = \frac{M}{2}$. This confirms the last row of Table 3.4.

Case 3: $n_1 > 1$ $(X = K2^{n_1}$; K **is an odd integer**).

Again, we write L_{s2} as

$$
\begin{aligned}
L_{s2} &= \left(\frac{2M}{\mathrm{GCD}(2M, (2|s_1[0] - K2^{n_1}|)(2^{n_0-n_1}))} \right) 2^{n_0-n_1} \\
&= \left(\frac{2M}{\mathrm{GCD}(2M, (|2s_1[0] - K2^{n_1}|)(2^{n_0-n_1}))} \right) 2^{n_0-n_1} \\
&= \left(\frac{2M}{\mathrm{GCD}(2M, (2|s_1[0] - K2^{n_1-1}|)(2^{n_0-n_1}))} \right) 2^{n_0-n_1} \\
&= \left(\frac{2M}{\mathrm{GCD}(2M, (|s_1[0] - K2^{n_1-1}|)(2^{n_0-n_1+1}))} \right) 2^{n_0-n_1}.
\end{aligned}
$$

In the last equation, the term $(|s_1[0] - K2^{n_1-1}|)$ is an odd integer. Thus, $\mathrm{GCD}(2M, (|s_1[0] - K2^{n_1-1}|)(2^{n_0-n_1+1})) = (2^{n_0-n_1+1})$. This yields $L_{s2} = 2^{n_0} = M$ and confirms the second row of Table 3.4.

In order to calculate L_{s3} using Table 3.6, we consider the following three cases and we use the three results for L_{s2}.

Case 1: $L_{s2} = 2M$.

Using Table 3.6, we can write L_{s3} as

$$
L_{s3} = \left(\frac{2M}{\mathrm{GCD}(2M, (|2s_2[0] - s_1[0]|)(2M))} \right) 2M.
$$

The denominator is equal to $2M$, giving $L_{s2} = 2M$.

Case 2: $L_{s2} = M$.

In this case, we have

$$
L_{s3} = \left(\frac{2M}{\mathrm{GCD}(2M, (|2s_2[0] - s_1[0]|)M)} \right) M.
$$

Since we assume an odd value for $s_1[0]$, the term $(|2s_2[0] - s_1[0]|)$ is an odd integer, giving $\mathrm{GCD}(2M, (|2s_2[0] - s_1[0]|)M) = M$. This results in

$$
L_{s3} = \frac{2M}{M} M = 2M.
$$

---

[16] Subtracting two odd integers results in an even integer.

Case 3: $L_{s2} = \frac{M}{2}$.

In this case, L_{s3} becomes,

$$L_{s3} = \left(\frac{2M}{\text{GCD}(2M, (|2s_2[0] - s_1[0]|)\frac{M}{2})} \right) \frac{M}{2}.$$

With similar reasoning, $\text{GCD}(2M, (|2s_2[0] - s_1[0]|)\frac{M}{2}) = (\frac{M}{2})$. This yields

$$L_{s3} = \frac{2M}{\frac{M}{2}} \frac{M}{2} = 2M.$$

Thus, we have proven that, provided the initial condition of the first stage $s_1[0]$ is an odd integer, the cycle length L_{s3} of the MASH 1-1-1 DDSM is $2M$, regardless of the input value X and the initial conditions of the second and third stages. We have repeated the simulation that yielded Fig. 3.27, but with a new set of initial conditions: $s_1[0] = 7$, $s_2[0] = 7$, and $s_3[0] = 0$. Figure 3.28 shows the result. As expected, the cycle length is always $2M$ since $s_1[0]$ has an odd value.

3.4.4 Using Prime Modulus Quantizers

In Sections 3.3.1 and 3.4.3, we studied in detail the effect on the cycle length of setting the initial condition. This technique belongs to the class of deterministic approaches to engineering the period of a DDSM. In this section, we study another method which is based on choosing special values for the quantizer modulus [15].

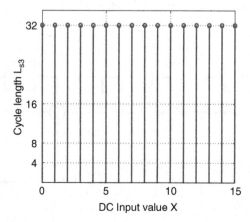

Fig. 3.28 Cycle length L_{s3} versus input in MASH 1-1-1 DDSM with $s_1[0] = 7$, $s_2[0] = 0$, $s_3[0] = 0$, and $M = 2^4$

Fig. 3.29 Input–output relationship of a 1-bit quantizer which thresholds at a prime integer M_p

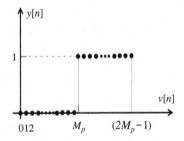

Consider a first order EFM with a 1-bit quantizer that thresholds at $M = M_p$, where M_p is prime (see Fig. 3.29). According to Table 3.6, the cycle length for this first order EFM is always M_p.

Based on this architecture, Level et al. [15] proposed a higher order MASH DDSM. Considering this system as a finite state machine and applying the methods of discrete-time dynamical systems, we will calculate its cycle length. We will prove that cascading stages of this type to build a higher order MASH does not increase the cycle length when the input is a constant. Rather, the cycle length is always equal to that defined by the first stage. Nevertheless, the advantage of the prime modulus MASH architecture is that it has a maximum guaranteed minimum cycle length for all constant inputs and for all initial conditions. Although stages after the first do not increase the cycle length, they do contribute to whitening the quantization noise.

The block diagram of such a MASH DDSM is shown in Fig. 3.30, where the EFM1 blocks use quantizers with a prime step size M_p; the corresponding transfer characteristic is shown in Fig. 3.29.

To analyze the prime MASH modulator in Fig. 3.30, we write the state equations where e_i is the state variable associated with stage i. Referring to Fig. 3.14a and Eq. (3.12), we write for the first stage of Fig. 3.30:

$$e_1[n] = v_1[n] \bmod M_p, \tag{3.85}$$

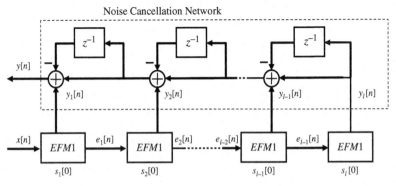

Fig. 3.30 Block diagram of the MASH DDSM. Each $EFM1$ Block has the structure shown in Fig. 3.14a with a prime modulus quantizer

where M_p is the modulus of the quantizer. Writing the equation at the input summing node, we obtain:

$$e_1[n] = (x[n] + e_1[n-1]) \bmod M_p, \tag{3.86}$$

where $x[n]$ is the input to the first stage. We can rewrite the above equation as:

$$e_1[n+1] = (x[n+1] + e_1[n]) \bmod M_p. \tag{3.87}$$

Repeating the same process for the ith stage, we obtain:

$$e_i[n+1] = (e_{i-1}[n+1] + e_i[n]) \bmod M_p, \tag{3.88}$$

where $e_{i-1}[n+1]$ is the error of the previous stage $(i-1)$ applied as the input to stage i. Starting from (3.87), assuming a constant input $x[n] = X$ (for all n), and expanding the indices of (3.88) from 1 to l, we derive a set of equations which we write in matrix form as:

$$E[n+1] = (AE[n] + BX) \bmod M_p, \tag{3.89}$$

where the lower triangular matrix A and the vectors B and E are defined by

$$A = \begin{pmatrix} 1 & 0 & \cdots & 0 & 0 \\ 1 & 1 & \cdots & 0 & 0 \\ \vdots & \vdots & & \vdots & \\ 1 & 1 & \cdots & 1 & 0 \\ 1 & 1 & \cdots & 1 & 1 \end{pmatrix}_{l \times l}, \tag{3.90}$$

$$B = \begin{pmatrix} 1 \\ 1 \\ 1 \\ \vdots \\ 1 \end{pmatrix}_{l \times 1} \quad \text{and } E = \begin{pmatrix} e_1 \\ e_2 \\ \vdots \\ e_l \end{pmatrix}_{l \times 1}. \tag{3.91}$$

Using (3.89) and expanding it over its indices from $E[n+2]$ to $E[n+N]$, we obtain:

$$
\begin{aligned}
E[n+2] &= (AE[n+1] + BX) \bmod M_p, \\
&= (A(AE[n] + BX) \bmod M_p + BX) \bmod M_p, \\
&= (A(AE[n] + BX) + BX) \bmod M_p, \\
&= (A^2 E[n] + (A^1 + A^0)BX) \bmod M_p.
\end{aligned} \tag{3.92}
$$

For index $n+3$, we write:

$$E[n+3] = (AE[n+2] + BX) \bmod M_p. \tag{3.93}$$

Substituting (3.92) into (3.93) and following the same process for achieving (3.92), we obtain:

$$E[n + 3] = \left(A^3 E[n] + \left(\sum_{k=0}^{2} A^k \right) BX \right) \bmod M_p. \tag{3.94}$$

Finally, for index $n + N$, we have:

$$E[n + N] = \left\{ A^N E[n] + \left(\sum_{k=0}^{N-1} A^k \right) BX \right\} \bmod M_p. \tag{3.95}$$

We substitute the following identities into (3.95):

$$A^N = \begin{bmatrix} 1 & 0 & \cdots & 0 & 0 \\ C_1^N & 1 & \cdots & 0 & 0 \\ \vdots & \vdots & & \vdots & \\ C_{l-2}^{N+l-3} & C_{l-3}^{N+l-4} & \cdots & 1 & 0 \\ C_{l-1}^{N+l-2} & C_{l-2}^{N+l-3} & \cdots & C_1^N & 1 \end{bmatrix}_{l \times l}, \tag{3.96}$$

$$\sum_{k=0}^{N-1} A^k = \begin{bmatrix} C_1^N & 0 & \cdots & 0 & 0 \\ C_2^N & C_1^N & \cdots & 0 & 0 \\ \vdots & \vdots & & \vdots & \\ C_{l-1}^{N+l-3} & C_{l-2}^{N+l-4} & \cdots & C_1^N & 0 \\ C_l^{N+l-2} & C_{l-1}^{N+l-3} & \cdots & C_2^N & C_1^N \end{bmatrix}_{l \times l}, \tag{3.97}$$

$$\left(\sum_{k=0}^{N-1} A^k \right) B = \begin{pmatrix} C_1^N \\ C_1^N + C_2^N \\ \vdots \\ C_1^N + C_2^N + \cdots + C_l^{N+l-2} \end{pmatrix}_{l \times 1} \tag{3.98}$$

$$= \begin{pmatrix} C_1^N \\ C_2^{N+1} \\ \vdots \\ C_l^{N+l-1} \end{pmatrix}_{l \times 1},$$

where $C_k^n = \frac{n!}{k!(n-k)!}$. In simplifying (3.98), we used Pascal's rule:

$$C_k^n + C_{k+1}^n = C_{k+1}^{n+1}. \tag{3.99}$$

Setting $n = 0$ in (3.95), we obtain

$$E[N] = \left(A^N E[0] + \left(\sum_{k=0}^{N-1} A^k\right) BX\right) \bmod M_p. \qquad (3.100)$$

If the system is periodic with period N, then we have that $E[N] = E[0]$; therefore, the following set of equations holds:

$$e_1[0] = \left(e_1[0] + C_1^N X\right) \bmod M_p,$$

$$e_2[0] = \left(C_1^N e_1[0] + e_2[0] + C_2^{N+1} X\right) \bmod M_p,$$

$$\vdots$$

$$e_l[0] = \left(C_{l-1}^{N+l-2} e_1[0] + C_{l-2}^{N+l-3} e_2[0]\right.$$

$$\left. + \cdots + e_l[0] + C_l^{N+l-1} X\right) \bmod M_p, \qquad (3.101)$$

where $e_i[0]$ is the initial error of the ith stage. If this time-invariant system starts from state $(e_1[0], \cdots, e_l[0])$ and returns to this state after N steps, then each signal e_i is periodic with period N. Therefore, in order for the above equations to be true, the following constraints must be satisfied:

$$(C_1^N X) = 0 \bmod M_p,$$

$$(C_1^N e_1[0] + C_2^{N+1} X) = 0 \bmod M_p,$$

$$\vdots$$

$$\left(C_{l-1}^{N+l-2} e_1[0] + C_{l-2}^{N+l-3} e_2[0] + \cdots + C_1^N e_{l-1}[0] + C_l^{N+l-1} X\right) = 0 \bmod M_p. \qquad (3.102)$$

We pick the first and the lth equations and write them as:

$$(NX) \bmod M_p = 0, \qquad (3.103)$$

$$N\left(\frac{(N+1)\cdots(N+l-2)}{(l-1)!} e_1[0] + \cdots + e_{l-1}[0]\right.$$

$$\left. + \frac{(N+1)\cdots(N+l-1)}{l!} X\right) \bmod M_p = 0. \qquad (3.104)$$

If we choose the modulus of the quantizers M_p to be prime, then the minimum integer solution for N in the first equation is $N = M_p$, with $0 < X < M_p$.

Now, consider the last term inside the parentheses in (3.104). We know that the product of l consecutive integers is divisible by $l!$. In other words, $l!$ divides $(N+1)(N+2)\cdots(N+l-1)(N+l)$. If N is a prime integer and $N > l$, $(N+l)$ is not divisible by l. Therefore, $l!$ must divide $(N+1)(N+2)\cdots(N+l-1)$; hence, $\frac{(N+1)\cdots(N+l-1)}{l!}$ is an integer. With a prime N and $N > l$, all the other terms

Table 3.7 Cycle lengths in a MASH DDSM with a prime M_p [15, 16]

Modulator order l	Worst case	Best case	Guaranteed minimum
$l > 1$	M_p	M_p	M_p

inside the outer parentheses of the second equation are integers for the same reason. Consequently, the solution of the lth stage is also $N = M_p$. In conclusion, if the quantizer modulus M_p is a prime number and if M_p is greater than the number of stages l, then the cycle length is always M_p, independently of the initial conditions $e_i[0]$, the value of the constant input X, and the order l of the MASH DDSM. This result is summarized in Table 3.7.

To illustrate the effect of this guaranteed minimum cycle length, we provide some examples. Figure 3.31 shows the spectra of two MASH 1-1-1 DDSMs, where $M = 512$ and $M_p = 509$, with the initial conditions of all stages set to zero, and with the input $X = 256$ in both cases. Note that the modulator with a prime modulus quantizer (M_p) distributes the quantization noise over significantly more tones, resulting from the longer cycle length compared to the case of just two tones when $M = 512$. Note that the cycle length in the prime MASH DDSM is always M_p, regardless of the input, the initial condition and the order. By contrast, the cycle length of the conventional MASH DDSM depends strongly on the input value, the initial condition and the order, as we have shown.

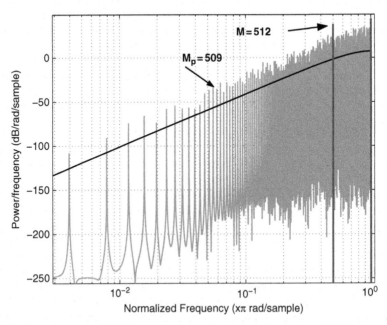

Fig. 3.31 Output spectra of MASH 1-1-1 with zero initial conditions and input 256 for two cases of $M = 512$ (power-of-two) and $M_p = 509$ (prime). The *solid curve* represents the expected spectrum, assuming additive white quantization noise. The number of samples taken for the spectral analysis is $N_f = 2^{18}$

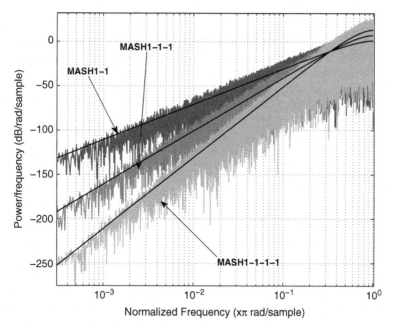

Fig. 3.32 Output spectra of MASH 1-1, MASH 1-1-1 and MASH 1-1-1-1 with input $X = \frac{2^{17}}{2} + \frac{2^{17}}{4} + \frac{2^{17}}{8}$. The quantizer step size is $M_p = 2^{17} - 1$ (prime). The *solid curve* represents the expected spectrum, assuming additive white quantization noise. The FFT length is $N_f = 2^{18}$

In order to obtain a smoother noise-shaped spectra, one can increase M_p. As an example, Fig. 3.32 shows the spectra of MASH 1-1, MASH 1-1-1, and MASH 1-1-1-1 modulators of the type shown in Fig. 3.30 with input $X = \frac{2^{17}}{2} + \frac{2^{17}}{4} + \frac{2^{17}}{8}$ when $M_p = 2^{17} - 1$ (prime).

The spectra are smooth and spike-free. The solid curves show the expected shaped noise spectra, assuming additive white quantization noise. Note that the spectrum of a second order modulator with an odd initial condition and with the same input is tonal (see Fig. 3.23) while the spectrum in this case is free of high power spurious tones; this is due to better randomization of the quantization error.

Figure 3.33 shows the autocorrelation functions of the quantization errors ($e_i[n]$) of the second, third and fourth stages in the MASH 1-1-1-1. The distances between the large spikes in the figure are equal to $2^{17} - 1$, which exactly matches the calculated periods of the quantizer error signals. In the case of the second stage, there are a few small spikes between each of the two large spikes, showing a small amount of correlation. In the case of the third and the fourth stages, for lags other than 0 and integer multiples of the period, the autocorrelation function is very small, indicating a very good approximation to white noise.

Thus, we can generate a spurious-tone free spectrum by choosing a prime M_p in higher order modulators (as suggested by the examples). Figure 3.34 shows interpolated output quantization noise spectra of a MASH 1-1-1, where $M_p = 2^{17} - 1$

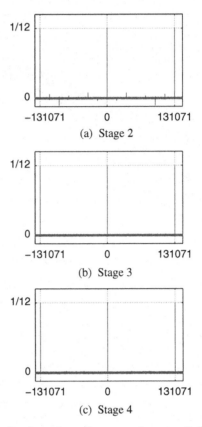

Fig. 3.33 Autocorrelation functions of quantizer errors for stages 2, 3 and 4 of MASH 1-1-1-1 with $M_p = 2^{17-1}$, $X = \frac{2^{17}}{2} + \frac{2^{17}}{4} + \frac{2^{17}}{8}$ and zero initial conditions

for inputs $X = 2^k$, $1 \le k \le 16$ and with zero initial conditions. Regardless of the input, the noise-shaped spectra are smooth and spike-free in this case.

3.5 Notes on MATLAB Simulations

Simulink models and MATLAB code are provided to enable the reader to reproduce the simulation results for each MATLAB plot given in this chapter. With the explanations for the MATLAB code and the Simulink models in Chapter 2, it should be straightforward to understand and run the codes related to this chapter. Therefore, we focus here on the parts which have not been explained earlier.

3.5.1 How to Calculate the Cycle Length

We use a function embedded in the Signal Processing Toolbox called "seqperiod". This function is in the form of

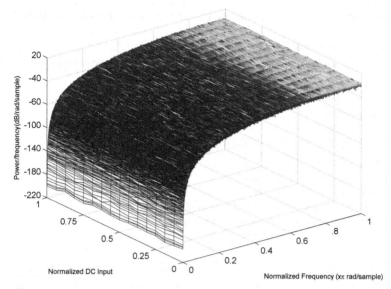

Fig. 3.34 Output quantization noise spectra of a MASH 1-1-1 for a range of DC inputs. The quantizer interval is $2^{17} - 1$ and all initial conditions are set to zero

$$p_x = \text{seqperiod}(x). \qquad (3.105)$$

It returns the integer p_x that corresponds to the period of the sequence in the vector x. The period p_x is computed as the minimum length of a subsequence $x(1:p_x)$ of x that repeats itself continuously every p_x samples in x. The length of x does not have to be a multiple of p_x, so that an incomplete repetition is permitted at the end of x. If x is not periodic, then $p_x = \text{length}(x)$. This function was used to generate Figs. 3.27 and 3.28.

Another method that can be used to measure the cycle length in a DDSM is by calculating the autocorrelation function [11]. As an example, for generating Fig. 3.33b, the following code is executed.

```
% Define the prime step size
n=17;
Mp=2^n-1;
% Define the number of simulation points
sim_time=2^18;
% Define the input
M=2^n;
input=M/2+M/4+M/8;
% Set the initial condition
ic1=0;
ic2=0;
ic3=0;
```

```
% Simulate the Simulink model for the prime modulus MASH 1-1-1
DDSM
sim('MASH111_prime.mdl',[1,sim_time];
% The error e3 is used for autocorrelation calculation using function
xcorr
xcorr_e3=xcorr((e3-mean(e3))/Mp,'unbiased');
figure; plot(xcorr_e3)
set(gca,'fontsize',12)
axis([2^17-10000 3*(2^17+5000) -0.01 1/10])
set(gca,'YTick',[0,1/12])
set(gca,'YTickLabel',{'0','1/12'} )
set(gca,'XTick',[2^(17)-1 2*(2^(17)-1) 3*(2^(17-1))])
set(gca,'XTickLabel',{'-131071','0','131071'})
set(findobj(gca,'Type','line'),'LineWidth',2)
grid on
```

This code uses the function "xcorr" to calculate the autocorrelation function. MATLAB calculates the *unbiased* cross-correlation of two signals x and y as follows:

$$c_{x,y}(l_i) = \begin{cases} \frac{1}{P-l_i} \sum_{n=0}^{P-l_i-1} x[n+m]y[n], & 0 \leq l_i \leq (P-1), \\ \frac{1}{P+l_i} \sum_{n=0}^{P+l_i-1} x[n]y[n+m], & -(P-1) \leq l_i < 0. \end{cases} \quad (3.106)$$

Here, P is the maximum lag specified by the user for the calculation, and l_i is the lag index. This formula is different from the conventional formula for the cross-correlation function since it has an additional multiplying factor $\frac{1}{P-l_i}$. This factor compensates for the finite duration of the two signals x and y and will compensate the autocorrelation estimate for lags close to P. The autocorrelation of a signal x is handled by assuming $x = y$ in Eq. (3.106). Hence, we obtain

$$c_{x,x}(m) = \frac{1}{P-m} \sum_{n=0}^{P-m-1} x[n+m]x[n], \quad -(P-1) \leq m \leq (P-1) \quad (3.107)$$

Note that we have removed the mean value of the error signal e_3, and have divided the result by M_p. In this way, the resulting signal is in the range $[-\frac{M_p-1}{2M_p}, \frac{M_p-1}{2M_p}]$. In MATLAB we have calculated

$$e_{3,\text{normalized}} = \frac{e_3 - \text{mean}(e_3)}{M_p},$$

and have applied e_3, normalized to the autocorrelation function.

The variance[17] of the resulting signal is $\frac{(M_p^2-1)}{12M_p^2} \approx \frac{1}{12}$, if the signal is uniformly distributed in the range $[-\frac{M_p-1}{2M_p} \ \frac{M_p-1}{2M_p}]$.

The autocorrelation function of a white noise signal has a nonzero value for $l_i = 0$ and is zero for $l_i \neq 0$. The value of the autocorrelation function at zero is equal to the variance of the signal and, if the white noise is uniformly distributed in the range $[-0.5 \ 0.5]$ (the case where M_p is large), its variance is $\frac{1}{12}$. Having a zero value of the autocorrelation function for all non-zero lags means that the samples of the white noise are completely uncorrelated with each other.

Ideally, we would like to have the autocorrelation function of the quantizer error of a DDSM approach that of white noise. However, we know that the quantizer error in the DDSM is periodic. The autocorrelation function of a periodic signal contains peaks that are separated by the period of the signal because, when the lag index is an integer multiple of the period, the shifted version of the signal is the same as the signal itself, giving the same value of autocorrelation as for zero lag. By randomizing the DDSM output using dithering or a deterministic technique, we aim to lengthen the period (pushing the peaks away from each other) and make the samples uncorrelated with each other within the periods. When the period is sufficiently long and the samples are sufficiently randomized, the autocorrelation function approaches that of white noise.

3.6 Summary

In this chapter, we have reviewed state-of-the-art techniques for maximizing cycle lengths in DDSMs. There are two classes of techniques for this purpose: stochastic and deterministic.

The most common method in the first category is called dithering. Dithering is effective in improving the modulator's tonal performance. However, it raises the noise floor in the spectrum of the output, particularly at low frequencies. Care should be taken to minimize this effect.

In the remainder of the chapter, we have studied deterministic techniques including seeding and using prime modulus quantizers. We showed empirically and mathematically that seeding helps to guarantee a minimum cycle length that is determined by the modulator word length n_0. In the case of higher order modulators, simulations show that one can achieve smooth spectra by choosing a large word length n_0. We also derived equations relating the cycle length to a given constant input and a set of initial conditions for first, second, and third order MASH DDSMs. Similar analyses have been performed for fourth and fifth order MASH DDSMs [72] and higher order EFMs [73].

Another method in the deterministic category is to use prime modulus quantizers in a MASH DDSM. We have shown mathematically that this idea guarantees a

---

[17] See Appendix A for calculation of the variance.

minimum cycle length M_p for an lth order MASH modulator, where M_p is the prime modulus value chosen for the quantizer. In summary, we report cycle lengths for conventional deterministic techniques in Table 3.8. In Table 3.9, we report cycle lengths in a MASH DDSM with a prime M_p.

In the next two chapters, we will study HK-MASH, HK-EFM and HK-SQ-DDSM modulators which improve on these results.

Table 3.8 Cycle lengths in conventional MASH DDSMs with a power-of-two modulus $M = 2^{n_0}$, [11, 13]

Modulator order l	Worst case $X = \frac{M}{2}$, $s_i[0] = 0$	Best case $s_1[0]$ odd	Guaranteed minimum $s_1[0]$ odd
2	4	$2M$	$\frac{M}{2}$
3	4	$2M$	$2M$
4	4	$4M$	$2M$
5	4	$4M$	$4M$

Table 3.9 Cycle lengths in a MASH DDSM with a prime modulus M_p [15, 16]

Modulator order l	Worst case	Best case	Guaranteed minimum
2	M_p	M_p	M_p
3	M_p	M_p	M_p
4	M_p	M_p	M_p
5	M_p	M_p	M_p

Chapter 4
Maximizing Cycle Lengths by Architecture Modification

4.1 Introduction

In Chapter 3, we studied conventional techniques for maximizing cycle lengths in DDSMs. We explained two classes of techniques, namely stochastic and deterministic approaches. In this chapter, we describe our deterministic technique for maximizing the cycle length by changing the architecture of the conventional MASH DDSM; the maximum cycle length structure is called HK-MASH.

In Chapter 5, we show how the technique described in this chapter can be applied to two other classes of modulators: EFM and SQ-DDSM.

In the first part of this chapter, we describe a modified first-order EFM. We then make use of the modified EFM to build the HK-MASH modulator. We study, by simulation, the spectral performance of second, third, and fourth order HK-MASH modulators with small and large accumulator word lengths. We conclude by describing an alternative cycle lengthening technique proposed by Song and Park [75].

4.2 Modified First Order Error Feedback Modulator

As mentioned in Chapter 3, the first-order EFM can be used as the building block in standard types of MASH DDSMs such as the MASH 1-1, MASH 1-1-1 and MASH 1-1-1-1. By adding an appropriate feedback path to the modulator, one can ensure that the cycle length achieves its theoretical maximum value for all inputs and for all initial conditions. A MASH DDSM with this modified EFM modulator yields a significantly greater cycle length compared to a seeded or a prime-modulus-quantizer MASH DDSM.

Figure 4.1 shows the modified architecture for the first-order EFM [76, 77]. This structure differs from the conventional EFM shown in Fig. 3.14a in that it includes a feedback block az^{-1}, where a is a specially-chosen small integer, to be discussed later. This modification yields the maximum cycle length[1] for all constant inputs, independently of the initial conditions.

---

[1] The cycle length is approximately equal to the number of available states.

K. Hosseini, M.P. Kennedy, *Minimizing Spurious Tones in Digital Delta-Sigma Modulators*, Analog Circuits and Signal Processing, DOI 10.1007/978-1-4614-0094-3_4, © Springer Science+Business Media, LLC 2011

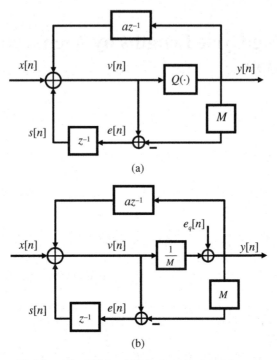

Fig. 4.1 (**a**) The modified first order EFM and (**b**) its linearized model

The 1-bit quantizer block $Q(\cdot)$ shown in Fig. 4.1 has the same input–output relationship as the conventional power-of-two quantizer shown in Fig. 3.14a, namely

$$y[n] = Q\big(v[n]\big) = \begin{cases} 0, & v[n] < M \\ 1, & v[n] \ge M, \end{cases} \tag{4.1}$$

where $M = 2^{n_0}$ and n_0 is the accumulator word length.

In Fig. 4.1b, the quantizer is modeled by the gain factor $\frac{1}{M}$ and an additive noise source (e_q) to investigate the overall operation of the structure in the z-domain.

We next derive an expression for the output y in terms of the input x and quantizer error e_q. In the z-domain,

$$Y(z) = \frac{1}{M} V(z) + E_q(z), \tag{4.2}$$

$$E(z) = V(z) - MY(z), \tag{4.3}$$

giving:

$$E(z) = -M E_q(z). \tag{4.4}$$

At the input summing node, we obtain:

$$V(z) = X(z) + az^{-1}Y(z) + z^{-1}E(z). \qquad (4.5)$$

Substituting (4.4) into (4.5), we obtain:

$$V(z) = X(z) + az^{-1}Y(z) - z^{-1}ME_q(z). \qquad (4.6)$$

Using the value of $V(z)$ from (4.2) and substituting into (4.6), we obtain:

$$MY(z) - ME_q(z) = X(z) + az^{-1}Y(z) - z^{-1}ME_q(z). \qquad (4.7)$$

Rearranging (4.7), we can write:

$$Y(z) = \frac{X(z)}{M - az^{-1}} + \frac{M(1 - z^{-1})}{M - az^{-1}}E_q(z) \qquad (4.8)$$

$$= \frac{1}{M}\frac{X(z)}{1 - \alpha z^{-1}} + \frac{(1 - z^{-1})}{1 - \alpha z^{-1}}E_q(z), \qquad (4.9)$$

where

$$\alpha = \frac{a}{M}. \qquad (4.10)$$

The signal and noise transfer functions are $\frac{1}{M}\frac{1}{1-\alpha z^{-1}}$ and $\frac{1-z^{-1}}{1-\alpha z^{-1}}$, respectively. Compared to a conventional first order digital EFM, a pole at $z = \alpha$ is added to both the signal transfer function (STF) and the noise transfer function (NTF). If α is sufficiently small, this pole is very close to the origin in the z plane; equivalently, it is a distant pole which does not significantly affect the overall operation of the modulator in the low frequency band.

What is special about this architecture compared to the conventional error feedback modulator? In Appendix B, we prove that the cycle length of the error signal e shown in Fig. 4.1a is:

$$N = M - a, \qquad (4.11)$$

for *all* constant inputs and for *all* initial conditions, where M is the switching threshold of the one-bit quantizer and a is determined from Table 4.1. For ease of implementation, we choose $M = 2^{n_0}$, where n_0 is the word length of the accumulator. a is chosen to make $(M - a)$ the largest prime number less than M. Note here that this technique differs from that described at the end of the last chapter [15, 16] where a prime modulus quantizer was used to produce a cycle length that is prime, in that a conventional power-of-two quantizer can be used in this case.

The cycle length of a first order EFM with a prime modulus quantizer is the same as the cycle length of the modified first order EFM when the chosen prime numbers M_p and $M - a$ are the same. In the latter case, implementing the quantizer is easier (since M is a power of 2) than in the case of the prime modulus quantizer used in [15, 16]. Table 4.1 shows the value a for n_0 in the range $5 \leq n_0 \leq 25$. The method can be extended to $n_0 > 25$ by choosing a as the difference between $M = 2^{n_0}$ and the largest prime number less than M.

Consider a word length of 17 ($n_0 = 17$). Table 4.1 specifies $a = 1$. In this case, the modified EFM structure has a cycle length of $2^{17} - 1$ for all constant inputs and for all initial conditions; this is approximately equal to the longest possible cycle length 2^{17}. Recall that the maximum period (2^{17}) in a conventional structure is only achievable for *some* combinations of digital inputs and initial conditions [11, 13]. In this structure, the maximum cycle length $(2^{17} - 1)$ is achieved in *all* cases.

Possible digital implementations of the modified EFM for the cases $a = 1$ and $a = 3$ are shown in Fig. 4.2a, b respectively. In Fig. 4.2a, a delayed version of the carry out signal is used as the additional feedback into the carry in. In the case of $a = 1$, this signal is simply applied to the carry-in input of the accumulator. For other values of a determined from Table 4.1, c is used to generate the binary equivalent of a. For example, in Fig. 4.2b, the two-bit signal applied to the adder assumes the values 0 or 3. When $c = 1$, a is added to the input; 0 is added to the input when $c = 0$.

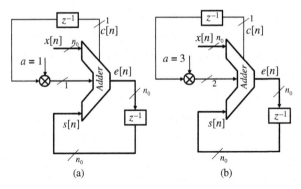

Fig. 4.2 Digital implementation of first order modified error feedback modulators with (**a**) $a = 1$ and (**b**) $a = 3$

Table 4.1 Corresponding values of a with respect to the modulator word length, for n_0 in the range 5–25 [see (4.11) and (4.18)]

n_0	a
5, 7, 13, 17, 19	1
6, 9, 10, 12, 14, 20, 22, 24	3
8, 18, 25	5
11, 21	9
16, 23	15
15	19

A conventional MASH structure can be built up from individual EFM stages. In a similar manner, we can build a modified MASH (denoted HK-MASH) using this modified first order modulator. We will show that an lth order HK-MASH has a guaranteed cycle length of $(M - a)^l$ for all inputs and for all initial conditions.

4.3 Maximized Cycle Length MASH Modulators (HK-MASH)

Using the structure introduced in the previous section, one can construct higher order MASH DDSMs [77], as shown in Fig. 4.3. In this figure, each block labeled MEFM1 contains the modified error feedback modulator structure shown in Fig. 4.1a. Using (4.8), one can calculate the z-domain input–output relationship. Note that $e[n] = -Me_q[n]$ in Fig. 4.1. Therefore,

$$Y_1(z) = \frac{1}{M} \frac{X(z)}{1 - \alpha z^{-1}} + \frac{1 - z^{-1}}{1 - \alpha z^{-1}} E_{q1}(z), \tag{4.12}$$

$$Y_2(z) = \frac{1}{M} \frac{-M E_{q1}(z)}{1 - \alpha z^{-1}} + \frac{1 - z^{-1}}{1 - \alpha z^{-1}} E_{q2}(z), \tag{4.13}$$

and, for the lth stage, we obtain:

$$Y_l(z) = \frac{1}{M} \frac{-M E_{q(l-1)}(z)}{1 - \alpha z^{-1}} + \frac{1 - z^{-1}}{1 - \alpha z^{-1}} E_{ql}(z). \tag{4.14}$$

Using the above equations, and considering Fig. 4.3, we write:

$$Y_M(z) = Y_1(z) + Y_2(z)(1 - z^{-1}) + \cdots + Y_{l-1}(z)(1 - z^{-1})^{l-1} \tag{4.15}$$

$$= \frac{1}{M} \frac{X(z)}{1 - \alpha z^{-1}} + \frac{(1 - z^{-1})^l}{1 - \alpha z^{-1}} E_{ql}(z), \tag{4.16}$$

and finally, we have:

$$Y(z) = \frac{1}{M} X(z) + (1 - z^{-1})^l E_{ql}(z), \tag{4.17}$$

where E_{ql} is the z-transform of the quantization error of the lth stage.

The signal and noise transfer functions of the modulator are the same as those of a conventional modulator. The HK-MASH DDSM differs from the conventional MASH DDSM in that it is constructed using the modified first-order EFM blocks shown in Fig. 4.1. It also has an additional output filter block denoted $(1 - \alpha z^{-1})$ in Fig. 4.3. Implementing this filter block could be difficult in practice; however, we do not need to implement the filter as the value of α is insignificant for a large value n_0. In fact, for a small value of α ($\alpha = \frac{a}{2^{n_0}}, n_0 > 5$), the spectra of y and y_M are approximately the same.

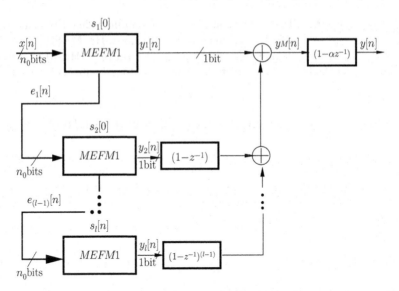

Fig. 4.3 Block diagram of a HK-MASH structure comprising cascaded modified first order EFMs described in the previous section. The MEFM1 block represents the modified EFM shown in Fig. 4.1

In Appendix B, we prove that the cycle length N of the HK-MASH structure shown in Fig. 4.3 is determined by the following equation:

$$N \equiv (2^{n_0} - a)^l, \tag{4.18}$$

where the value of a is determined from Table 4.1. For example, in the case of $n_0 = 19$ and $l = 3$, referring to Table 4.2 and (4.18), the cycle length is $(2^{19} - 1)^3$ for *all* constant digital inputs and for *all* initial conditions. We reiterate that these results are *always* achieved, without seeding or dithering.

In Table 4.2, we compare these results with the results for sample deterministic strategies described in [11, 13, 15, 16]. We have assumed a prime number $M_p = (2^{n_0} - a)$ for the switching threshold value of the 1-bit quantizer in [15, 16].

Table 4.2 A comparison of guaranteed minimum cycle lengths of the HK-MASH DDSM (based on power-of-two modulus quantizers) with seeded MASH DDSM (based on power-of-two quantizers) and prime-modulus-quantizer DDSM

Order l	[11, 13] power-of-two modulus odd initial condition (i.c.)	[15, 16] prime modulus $M_p = (2^{n_0} - a)$ *all* i.c. and constant input	HK-MASH power-of-two modulus *all* i.c. and constant inputs
2	$2^{(n_0-1)}$	$(2^{n_0} - a)$	$(2^{n_0} - a)^2 \approx (2^{2n_0})$
3	$2^{(n_0+1)}$	$(2^{n_0} - a)$	$(2^{n_0} - a)^3 \approx (2^{3n_0})$
4	$2^{(n_0+1)}$	$(2^{n_0} - a)$	$(2^{n_0} - a)^4 \approx (2^{4n_0})$
5	$2^{(n_0+2)}$	$(2^{n_0} - a)$	$(2^{n_0} - a)^5 \approx (2^{5n_0})$

Note that adding a stage to the HK-MASH structure increases the cycle length by approximately 2^{n_0} compared to a factor of no more than 4 if the conventional seeding technique is used. As proven in the previous chapter, the cycle length of the prime modulus quantizer remains the same, independently of an increase in the order of the MASH structure.

Since α is small, the error introduced by omitting the filter $(1 - \alpha z^{-1})$ is negligible. Therefore, in the following section, we consider the signal $y_M[n]$ rather than $y[n]$, assuming that the filter $(1 - \alpha z^{-1})$ is not implemented.

4.4   Performance Comparison of Different Maximized MASH DDSMs

4.4.1   Spectral Investigation

In order to illustrate the effect of randomization achieved by maximizing the cycle length for high order MASH modulators, we present the associated power spectra and the autocorrelation function of the quantizer error in the last stage. Simulations have been performed with modulator orders $l = 2, 3,$ and 4 for four cases; (a) prime modulus, (b) seeded, (c) dithered, and (d) HK MASH DDSMs.

4.4.1.1   Modulator Order $l = 2$

The input to all modulators is set to $X = \frac{M}{2} + \frac{M}{4} + \frac{M}{8}$, where $M = 2^{13}$. According to Table 4.1, $a = 1$ is chosen for the HK-MASH DDSM. The step-size of the prime-modulus modulator is $M_p = 2^{13} - 1$ (prime). The initial conditions of all stages are set to zero, except in the case of the seeded-MASH DDSM, where $s_1[0] = 53$ (odd) and $s_2[0] = 0$. A nonshaped LSB dithering signal d is added to the input in the case of the dithered-MASH DDSM. The corresponding cycle lengths of the deterministic modulators are 2^{13}, $2^{13} - 1$ and $(2^{13} - 1)^2$ for the seeded, prime-modulus, and HK modulators, respectively (as proven in Chapter 3 and Appendix B). Figure 4.4 shows pairwise the HK-MASH spectrum and the spectra of the other three modulators for ease of comparison.[2]

Referring to Fig. 4.4a, we see that the spectrum of the HK-MASH DDSM is closer to the solid curve which represents the idealized-shaped-white quantization noise. This is due to the longer cycle length of the HK-MASH modulator compared to the prime-modulus MASH DDSM.

Figure 4.4b shows spectral plots of (i) the HK-MASH DDSM, and (ii) the seeded-MASH DDSM. The spectrum of the seeded-MASH modulator, deviates significantly from the solid curve at high frequencies. In this case, the cycle length of the seeded MASH is shorter than the cycle length of the HK-MASH DDSM and,

---

[2] In the spectral plots presented in this chapter, the solid curves represent the idealized shaped white quantization noise, including the dither contribution where applicable.

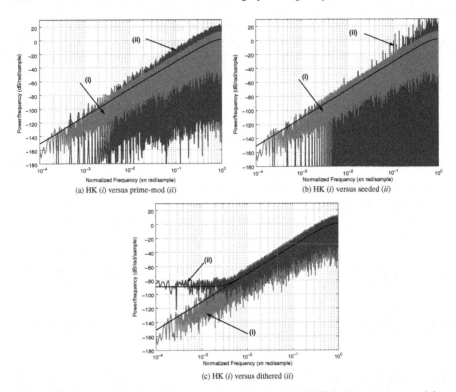

Fig. 4.4 Comparisons of the output spectra of the HK ($a = 1$) DDSM with (**a**) prime-modulus ($M_p = 2^{13} - 1$), (**b**) seeded ($s_1[0] = 53, s_2[0] = 0$), and (**c**) dithered ($R = 0$, see Table 3.1) MASH 1-1 DDSMs. In all cases, $X = \frac{M}{2} + \frac{M}{4} + \frac{M}{8}$ where $M = 2^{13}$, and the number of output samples is 2^{18}. The initial conditions of all modulators, except the seeded MASH 1-1, are zero

as we will demonstrate by studying the autocorrelation function, the seeded-MASH 1-1 DDSM fails to randomize the quantization noise effectively.

Comparing the spectra of the seeded and prime-modulus MASH DDSMs (Fig. 4.4a, b), we conclude that the prime-modulus modulator does a better job of randomization, resulting in a better spectrum, although both have similar cycle lengths.

Finally, we have plotted the spectra of the HK and the dithered MASH DDSMs in Fig. 4.4c. The maximum allowed order R of the dither filter is 0 (see Table 3.1). The nonshaped dither signal randomizes the modulator, resulting in a tone-free spectrum; however, it raises the noise floor to $-90\,dB$.

In order to show the effect of the randomization in the time domain, we have plotted in Fig. 4.5 the autocorrelation function for the quantizer error of the last stage in each of the above modulators.[3]

---

[3] The details of calculating the autocorrelation function in MATLAB are given in Section 3.5.

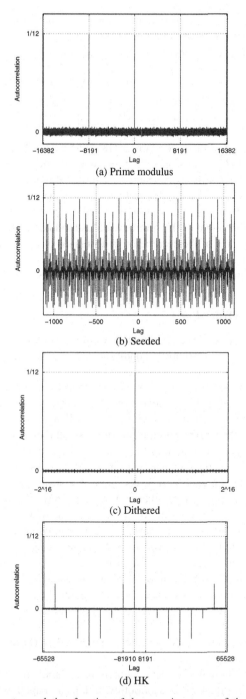

Fig. 4.5 Calculated autocorrelation function of the quantizer error of the second stage in the MASH 1-1 DDSM for four configurations: (**a**) prime modulus, (**b**) seeded, (**c**) dithered, and (**d**) HK. $M = 2^{13}$, $M_p = M - a$, $X = \frac{M}{2} + \frac{M}{4} + \frac{M}{8}$. The nonzero seed value of the seeded MASH DDSM is $s_1[0] = 53$

As seen in Fig. 4.5, the autocorrelation of the prime-modulus MASH DDSM has a train of peaks located at integer multiples of the period $2^{13}-1$. By comparison with the seeded-MASH DDSM [plot (b)], the prime-modulus MASH DDSM randomizes the quantizer error of the last stage more effectively; that is why one obtains better spectral performance with the prime-modulus MASH DDSM. The autocorrelation function of the dithered-MASH DDSM has only one large peak at the zero lag and the value of the autocorrelation function is small (0.0027) for non-zero lags, resembling the autocorrelation function of a white noise process.

Although the period of the second order HK-MASH DDSM is $(M-a)^2 \approx 2^{26}$, there is strong periodic behavior [structures with shorter periods, as shown in Fig. 4.5d]. This degrades the performance of the modulator so that it looks as if the period is only $M-a$ (8191); the first peak (other than the one at the zero lag) is located at 8191. With this observation, we would expect that the spectrum of the HK-MASH is similar to that of the prime-modulus modulator, perhaps with a slight improvement, as the size of the correlation is decreased by increasing the lag index. This is confirmed by the spectral plots shown in Fig. 4.4.

Next, we will compare the performance of the HK-MASH modulator with $l=3$.

4.4.1.2 Modulator Order $l=3$

Modulators of order $l=3$ were simulated. In this case, $n_0 = 9$, $M = 2^9$, $X = \frac{M}{2} + \frac{M}{4} + \frac{M}{8}$, $a = 3$ and zero initial conditions are imposed in all cases except for the seeded-modulator for which $s_1[0] = 53$, $s_2[0] = 0$, and $s_3[0] = 0$. The step-size of the prime-modulus quantizer is $M-a$ in the case of the prime-modulus MASH DDSM. The value of M in this simulation has been reduced in order to illustrate the performance of different modulators with short wordlengths.

Figure 4.6 shows the autocorrelation functions for all four cases. From Chapter 3, we recognize that the cycle length of the prime-modulus modulator is $M-a$. Plot (a) suggests the same, as the distance between the two peaks having the same value is 509 samples. Between each two peaks, there is a strong correlation among the samples compared to that of the dithered MASH DDSM. The cycle length of the seeded modulator is 1024 according to our previous calculations in Chapter 3. Plot (b) suggests the same, where the distance between each two peaks having the same value is 1024. At odd integer multiples of 512, between each two large peaks, there is a strong correlation, with a value of approximately -0.042. The autocorrelation functions of the dithered and HK modulators are shown in Fig. 4.6c, d for a long lag (2^{17}). In both cases, there is only a peak at zero lag and the autocorrelation function has small values for non-zero lags. Therefore, for both cases, we expect to obtain tone-free spectra.

The corresponding power spectra are shown in Fig. 4.7. These confirm our observations related to the autocorrelation function.

As suggested by the autocorrelation function, the spectrum of the HK-MASH modulator (shown in Fig. 4.7) (i) is smoothly high pass shaped, while the spectrum of the prime-modulus quantizer (Fig. 4.7a) (ii) contains a number of undesirable large tones that rise significantly above the solid curve.

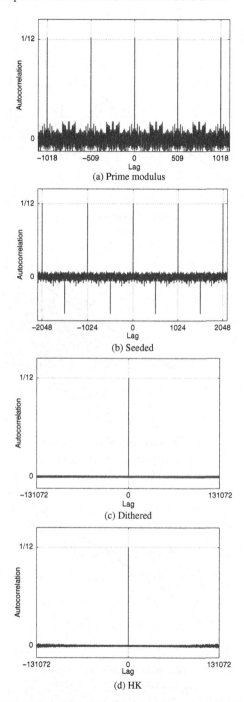

Fig. 4.6 Calculated autocorrelation function of the quantizer error of the third stage in a MASH 1-1-1 DDSM for four configurations: (**a**) prime modulus, (**b**) seeded, (**c**) dithered, and (**d**) HK. $M = 2^9, a = 3, X = \frac{M}{2} + \frac{\tilde{M}}{4} + \frac{M}{8}, M_p = 2^9 - 3$. The seed value of the seeded MASH DDSM is $s_1[0] = 53$. The initial states of the other modulators were set to zero

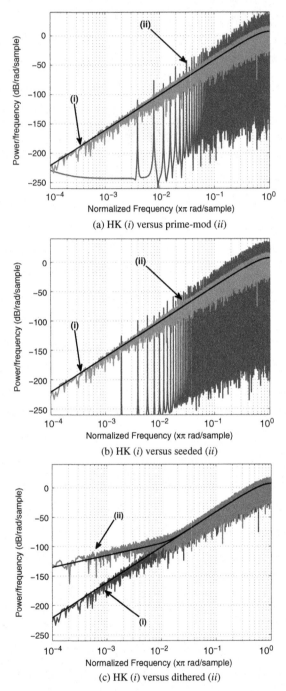

Fig. 4.7 Output spectra of the 9-bit HK ($a = 3$) DDSM with (**a**) prime-modulus ($M = 2^9 - 3$), (**b**) seeded ($s_1[0] = 53$, $s_2[0] = 0$, $s_3[0] = 0$), and (**c**) dithered ($R = 1$, see Table 3.1) MASH 1-1-1 DDSMs. In all cases, $X = \frac{M}{2} + \frac{M}{4} + \frac{M}{8}$ where $M = 2^9$ and the number of output samples for analysis is 2^{18}. The initial conditions of all modulators, other than the seeded one, are zero

Figure 4.7b compares the spectrum of the HK-MASH modulator with that of the seeded MASH DDSM. Similar to the case of the prime-modulus modulator, the seeded MASH modulator has a short cycle with samples correlated between the cycles. This results in a strongly tonal spectrum, as shown in the figure.

Finally, we have plotted in Fig. 4.7c the spectra of the dithered MASH modulator and the HK-MASH modulator. As expected from the autocorrelation functions, both have smooth spectra; however, the dithered MASH modulator has an elevated noise floor as before, albeit first order shaped.

Next, we illustrate the performance of the fourth order MASH DDSM ($l = 4$).

4.4.1.3 Modulator Order $l = 4$

Figure 4.8 shows the autocorrelation functions of the four types of MASH modulator considered as before, this time with $M = 2^7$ and $M_p = 2^7 - 1$.

Again, the prime-modulus and the seeded MASH modulators have very short cycles (127 and 256 respectively); therefore, we expect their spectra to contain a small number of large tones. The autocorrelation functions of the HK and dithered MASH DDSM are similar to that of white noise; therefore, one would expect to obtain smooth spectra.

The spectral plots shown in Fig. 4.9 are consistent with the autocorrelation functions, namely that, for the case of the seeded and prime-modulus modulators, the spectra contain undesirable high power tones, while the spectra of the dithered and HK modulators are smoothly shaped. The slope of the spectrum is 80 dB/decade in each case, as expected. In the case of the dithered modulator, we have applied an additive second-order-shaped LSB dither signal to the input; this reduces the slope to 40 dB/decade.

4.4.2 Experimental Results

Figure 4.10 shows experimentally measured results for (a) a seeded MASH 1-1-1 modulator with $M = 2^9$, $s_1[0] = 1$, $s_2[0] = s_3[0] = 0$ and $X = 256$, (b) a seeded MASH 1-1-1 modulator with $M = 2^{18}$, $s_1[0] = 1$, $s_2[0] = s_3[0] = 0$ and $X = 2^{17}$, and (c) an HK-MASH 1-1-1 DDSM with $M = 2^9$, $s_1[0] = s_2[0] = s_3[0] = 0$, $a = 3$ and $X = 256$. The modulators were implemented on a Xilinx Spartan II FPGA [78]. The digital outputs in both cases are applied to an Analog Devices Eval-AD5424 DAC board [79]. The spectra were measured using an Agilent EE4402B spectrum analyzer.

The measured results confirm qualitatively the theory and simulations. The measured spectrum of the 9-bit MASH in Fig. 4.10 contains a large number of discrete tones. These result from the short non-white quantization noise sequence. As expected, the increase in the cycle length in the 9-bit HK-MASH distributes the quantization noise power over many more frequency components. Consequently, the measured spectrum is much smoother over the entire frequency range. The spectra of the 18-bit MASH and 9-bit HK-MASH are almost indistinguishable.

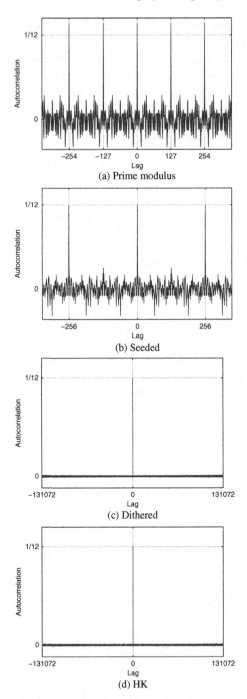

Fig. 4.8 Calculated autocorrelation functions of the quantizer error of the fourth stage in four MASH 1-1-1-1 DDSM configurations: (**a**) prime modulus ($M-1$), (**b**) seeded ($s_1[0] = 53$, $s_2[0] = 0$, $s_3[0] = 0$, $s_4[0] = 0$), (**c**) dithered (second order shaped), and (**d**) HK. $M = 2^7$ and $X = \frac{M}{2} + \frac{M}{4} + \frac{M}{8}$

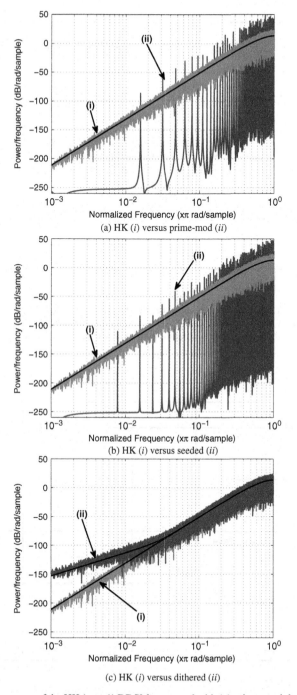

Fig. 4.9 Output spectra of the HK ($a = 1$) DDSM compared with (**a**) prime-modulus ($M = 2^7 - 1$), (**b**) seeded ($s_1[0] = 53$, $s_2[0] = 0$, $s_3[0] = 0$, $s_4[0] = 0$), and (**c**) dithered ($R = 2$, see Table 3.1) MASH 1-1-1-1 DDSMs. In all cases, $X = \frac{M}{2} + \frac{M}{4} + \frac{M}{8}$ where $M = 2^7$ and the number of output samples for analysis is 2^{18}. The initial conditions of all modulators other than the seeded one, are zero

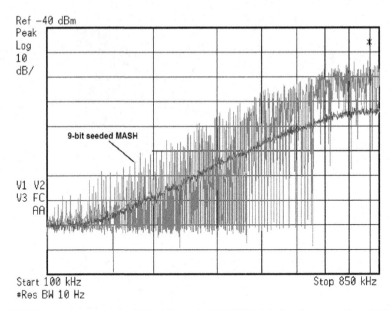

Fig. 4.10 Measured spectra of a 9-bit odd seeded MASH 1-1-1 (tonal spectrum), a 9-bit HK-MASH 1-1-1 with $X = 256$, with $X = 256$ and an 18-bit odd seeded MASH 1-1-1 with $X = 2^{17}$. The output spectra of the 18-bit MASH and 9-bit HK-MASH are almost indistinguishable

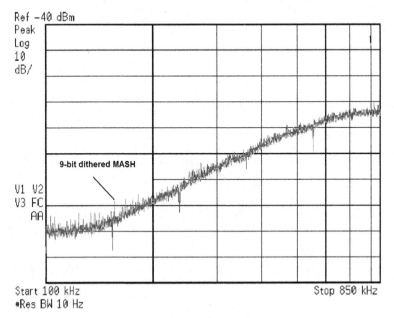

Fig. 4.11 Spectra of a 9-bit third order HK-MASH 1-1-1, a 9-bit shaped dithered MASH 1-1-1 and an 18-bit shaped dithered MASH 1-1-1 with input $X = 1$

Figure 4.11 shows measured results for a 9-bit first-order shaped LSB dithered MASH 1-1-1, an 18-bit first order shaped LSB-dithered MASH 1-1-1 and a 9-bit HK-MASH 1-1-1 with input $X = 1$ in all the cases. The spectra are almost smooth and matched over the range shown, except for some small low frequency spurs on the spectrum of the 9-bit dithered MASH 1-1-1. This experiment confirms that the spectral performance of the 9-bit HK-MASH is as good as that of an 18-bit MASH, and better at low frequencies than the dithered 9-bit MASH.

The low frequency sections of the spectra are flattened due to the noise floor of the measuring instrument so it is not possible to draw conclusions from this experiment concerning the difference predicted in Fig. 4.7c.

4.4.3 Relative Hardware Complexity

We applied the 1-bit output of a 21-bit linear feedback shift register (LFSR) to the carry-in of the input adder of the second stage of the MASH 1-1-1 DDSM. The LFSR circuit that generates the 1-bit dither signal is shown in Fig. 4.12. It uses the

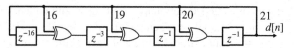

Fig. 4.12 Circuit diagram of a 21-bit LFSR circuit used for experimental evaluation

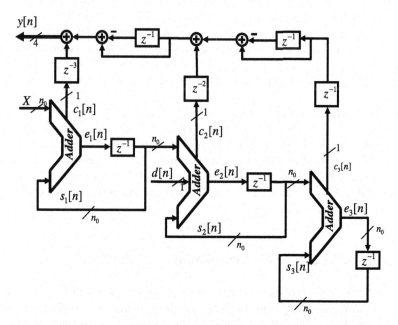

Fig. 4.13 Circuit diagram of the dithered MASH 1-1-1 DDSM used for the experimental evaluation. The input to the modulator is X, the output is y and the dither signal is d

Table 4.3 Hardware comparison of first order shaped dithered MASH 1-1-1 and HK-MASH 1-1-1. The numbers in the 3rd and 4th columns are the total equivalent gate counts

Word-length	a	Dithered MASH	HK-MASH	$\dfrac{\text{HK-MASH}}{\text{Dithered MASH}}\%$
9	3	1102	1091	99
11	9	1270	1295	102
13	1	1438	1238	86
15	19	1606	1721	107
17	1	1774	1574	89
19	1	1942	1742	90
24	3	2362	2621	111

classical Galois implementation [80] and the taps making the related polynomial primitive were chosen to be 21, 20, 19 and 16; the output is fed back to the XOR gates located at these taps. Note that there are several solutions to maximize the cycle length of the LFSR circuit by choosing different taps but all yield the same cycle length of $2^{21} - 1$ [80]. Here, we have arbitrarily chosen one such solution.

The block diagram of the dithered modulator is shown in Fig. 4.13. The dither signal d is applied to the carry in of the second stage. This ensures a compact realization that is equivalent to first-order-shaped LSB dithering added to the input. In this case, there is no need to apply the generated dither signal to an additional filter $(1 - z^{-1})$ nor to instantiate an extra adder to add the resulting signed shaped dither signal at the input of the modulator.

For the HK-MASH 1-1-1 structure, the implementation is simplest when $a = 1$, as the 1-bit feedback signal from the output of the block az^{-1} in Fig. 4.1 is simply applied to the carry-in of the input adder. One approach for the HK-MASH DDSM when $a \neq 1$ is to implement another adder of size $n_0 + 1$ in each stage, where the sum of $x[n]$ and $s[n]$ is applied to one input and the 2- to 5-bit word from the output of the block az^{-1} is applied to the LSBs of the other input (whose MSBs are set to zero).

Sample results are reported in Table 4.3 for the modulators described above with word-lengths 7–25 synthesized using the Xilinx ISE 8.2i tool. For word-lengths 13, 17 and 19 ($a = 1$), the HK-MASH requires less hardware than the dithered-MASH. The HK-MASH DDSM consumes almost the same amount of hardware as the dithered-MASH for word-lengths 9 and 11. For word-lengths 15 and 24, the dithered MASH has 7 and 11% less hardware, respectively, when compared to 15- and 24-bit HK-MASH DDSMs.

4.5 Song and Park Cycle Lengthening Architecture

Song and Park have proposed an alternative modified MASH DDSM, shown in Fig. 4.14, which also lengthens the cycle length by architecture modification. The first stage uses the conventional EFM1 configuration. The subsequent stages are constructed based on a modified first order EFM configuration (denoted $EFM1P$).

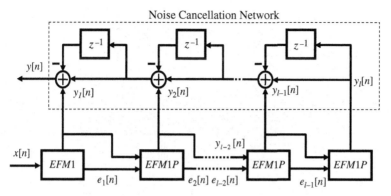

Fig. 4.14 Block diagram of the MASH DDSM proposed in [75]. The first stage uses the conventional EFM1 (denoted EFM1C in the figure) and the subsequent stages use the EFM stage (denoted EFM1P) shown in Fig. 4.15

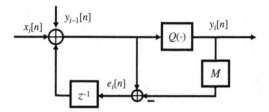

Fig. 4.15 Block diagram of the EFM stage (denoted EFM1P) used in Fig. 4.14

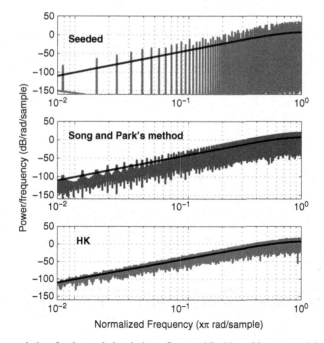

Fig. 4.16 Spectral plots for the seeded technique, Song and Park's architecture and the HK-MASH DDSM

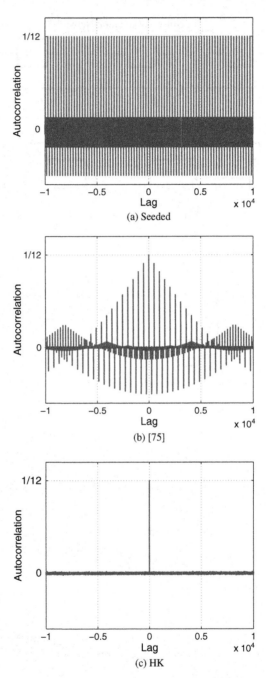

Fig. 4.17 Calculated autocorrelation functions of the quantizer error of the third stage in three MASH 1-1-1 DDSM configurations: (**a**) Seeded ($s_1[0] = 1$) (**b**) Song and Park (**c**) HK-MASH. $M = 2^7$ and $X = \frac{M}{2}$

The modified EFM1 shown in Fig. 4.15 is almost the same as the conventional EFM1 but has an extra input which is fed by the 1-bit output of the previous stage. If the cycle length of the first stage is N_1, then the cycle length of a MASH 1-1-1 DDSM based on this technique is $N_1 M^2$ where $M = 2^{n_0}$ [75]. N_1 depends on the input; it assumes a minimum value of 2 and a maximum value of M. Therefore, the minimum cycle length is $2M^2$ and the maximum value is M^3. The cycle length achieved by this technique is larger than that of the seeding technique but its guaranteed value is smaller than that of the HK-MASH DDSM architecture.

In Fig. 4.16 we show simulation results for the seeded MASH DDSM, Song and Park's technique [75] and for the HK-MASH DDSM with the following parameters: the orders of the modulators are three; $n_0 = 7$, $X = \frac{M}{2}$; NFFT $= 2^{16}$. An odd initial condition ($s_1[0] = 1$) is used for the seeded technique and zero initial condition for the other two techniques. In this example, Song and Park's technique [75] increases the cycle length and randomizes the quantization noise, leading to a smoother spectrum when compared to the plot of the seeding technique. However, the HK-MASH DDSM has a smoother spectrum and has a smaller (by 5 dB) spectral envelope at higher frequencies due to its better randomization. This is confirmed by the autocorrelation plots in Fig. 4.17 which show that the HK-MASH DDSM is doing a better job of randomization comparing the other two techniques.

4.6 Summary

In this chapter, we described a modified first order error feedback modulator which yields the maximum cycle length for *all* constant inputs and for *all* initial conditions. The modified MASH DDSM based on this structure (HK-MASH) has a significantly increased cycle length for all DC inputs and for all initial conditions compared to a conventional undithered MASH. The cycle length is approximately $2^{n_0 l}$, where n_0 is the word-length and l is the number of first order stages.

Simulations, confirmed by experiments, show the effect of maximized cycle lengths on the noise shaping spectra of these modulators. Simulations suggest that for the HK-MASH 1-1-1 DDSM, word-lengths as small as 9 can be used to realize smooth spectra. We also presented briefly an alternative deterministic technique for maximizing the cycle length.

The HK-MASH DDSM provides a deterministic approach that compares favorably with state of the art dithering strategies for maximizing cycle lengths. While consuming almost the same amount of hardware, it has better spectral performance at low frequencies.

Chapter 5
HK-EFM and HK-SQ-DDSM

5.1 Introduction

The last two chapters addressed techniques for maximizing cycle lengths in MASH DDSMs. Chapter 3 presented an overview of conventional stochastic and deterministic methods. Chapter 4 described a deterministic technique [76] for maximizing cycle lengths in MASH DDSMs.

In this chapter, we show how the idea described in Chapter 4 can be extended to two classes of multi-bit modulators [81]: higher order Error Feedback Modulators (EFMs) and single-quantizer DDSMs (SQ-DDSMs). After presenting background material on EFMs, we show spectral simulation results for maximum cycle length multi-bit EFMs. In Section 5.4, we present simulation results for maximum cycle length SQ-DDSMs. Implementation issues are discussed in Section 5.5.

5.2 HK-EFM

The results described in Chapter 4 guarantee a maximized minimum cycle length for *all* constant inputs and for *all* initial conditions in the case of MASH DDSMs incorporating first order EFMs with a 1-bit quantizer. In order to achieve this performance, the cycle lengths of constituent EFMs are maximized by introducing a feedback path from the output of each EFM to its input using a delay block with a particular choice of coefficient a, where a is a small integer that is chosen such that $M - a$ is prime.

The length of the output sequence in this modulator has been proven mathematically to be equal to $(M - a)^l$, where l is the order of the MASH modulator (equal to the number of the first order stages). When used in a high order MASH DDSM, this technique has been shown empirically to outperform conventional LSB dithering [9] in terms of its spectral performance.

In this chapter, we show how the same idea can be applied to higher order EFMs and higher order SQ-DDSMs. First, we consider EFMs. The generic block diagram of a digital multi-bit EFM is shown in Fig. 5.1. The modulator contains an optional additive input dither signal d, an optional filter (denoted $(1 - z^{-1})^R$) to shape the

K. Hosseini, M.P. Kennedy, *Minimizing Spurious Tones in Digital Delta-Sigma Modulators*, Analog Circuits and Signal Processing, DOI 10.1007/978-1-4614-0094-3_5, © Springer Science+Business Media, LLC 2011

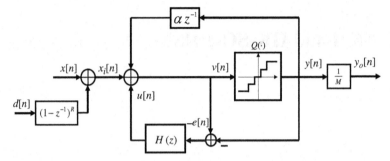

Fig. 5.1 Generic block diagram of a higher order EFM with a multilevel quantizer, shaped additive input LSB dither, and an output feedback path αz^{-1} which deterministically maximizes the number of spectral tones

dither, and an output feedback path (denoted αz^{-1}) which will be used to maximize the cycle length in a deterministic way. The multi-level quantizer [denoted $Q(\cdot)$] provides a coarse approximation y of the digital signal v.

Figure 5.2 shows the input–output characteristic of the multi-level mid-tread quantizer which we consider for our simulations. Each quantization step is of length M, there are $(N_{\max} + N_{\min} + 1)$ output levels, and the no-overload range is defined by $-(N_{\min}M + M/2) \leq v \leq (N_{\max}M + M/2 - 1)$. For ease of implementation, we assume that the step size M of the quantizer is a power of two ($M = 2^{n_0}$). The EFM is one possible realization of the noise shaping technique in the digital domain [4].

Choosing $H(z) = 1 - (1 - z^{-1})^l$, the signal transfer function $STF(z) = \frac{Y(z)}{X_i(z)}$ and noise transfer function $NTF(z) = \frac{Y(z)}{E(z)}$ become

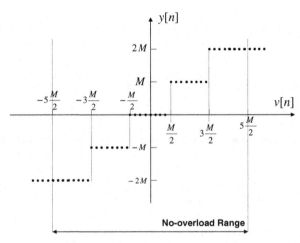

Fig. 5.2 The input–output characteristic of the multi-level quantizer with $N_{\min} = 2$ and $N_{\max} = 2$

$$STF(z) = \frac{1}{1 - \alpha z^{-1}}; \quad NTF(z) = \frac{(1 - z^{-1})^l}{1 - \alpha z^{-1}}. \tag{5.1}$$

Next, we split the discussion into two sub-sections: the conventional architecture and the HK architecture.

5.2.1 Conventional Architecture

The case $\alpha = 0$ corresponds to a conventional DDSM, where the EFM has an all-pass signal transfer function $STF(z) = 1$ and a high-pass noise transfer function $NTF(z) = (1 - z^{-1})^l$ that rejects the quantization noise at low frequencies. l determines the order of the modulator. The larger the value of l, the more the quantization noise is attenuated at low frequencies. The quantizer error $e = y - v$ is usually assumed to be white noise. With this assumption, the DDSM spectrum is smooth, and the noise is concentrated toward high frequencies, away from the signal band.

Like the MASH DDSM, the digital EFM is a Finite-State Machine (FSM) which always produces a *periodic* output signal (a cycle) when the input is a constant. In this case, the quantization noise signal is also periodic. In general, the period depends on the input, the initial conditions, and the order of the modulator. When the period is short, the power of the sequence is distributed among a limited number of undesirable tones (so-called "spurious tones") that appear in the DDSM output spectrum. The powers of the spurious tones (spurs) can be significantly higher than the noise-shaping curve predicted by the simplifying assumption that the quantizer can be modelled as an additive white noise source.

Figure 5.3 also illustrates the problem of short cycle lengths in EFMs. It shows MATLAB simulation results for a third order implementation of the EFM shown in Fig. 5.4 ($\alpha = 0$ and $d[n] = 0 \forall n$). The quantizer step size is $M = 2^{17}$, $N_{\min} = 4$, $N_{\max} = 4$, the filter is third order ($l = 3$), and all initial conditions are set to zero. All simulations have been performed using signed integer arithmetic.

Plot (i) shows the simulation result when the input is 1. Plot (ii) shows the spectrum of the output sequence when $x = 2^{16}$. Note that plot (i) can be approximated by a smooth curve, shown solid, which results from assuming that the quantizer adds uniformly-distributed white noise. By contrast, the spectrum shown in plot (ii) contains only two high power tones because the quantization noise is far from white. This example shows how the modulator can fail catastrophically to perform proper noise shaping, depending on the input value.

Using a stochastic approach, one can apply a pseudo-random binary dither sequence d (shown in Fig. 5.1) to the input [9, 61]. This breaks up the cycles and increases the effective cycle length, resulting in a smoother noise-shaped spectrum. While it increases the cycle length, as required, dithering inherently adds noise to the spectrum; care must be taken to minimize the contribution of this additional noise. An alternative (deterministic) approach is to avoid known short cycles by setting the initial conditions of the internal registers of the EFM [11], as discussed in

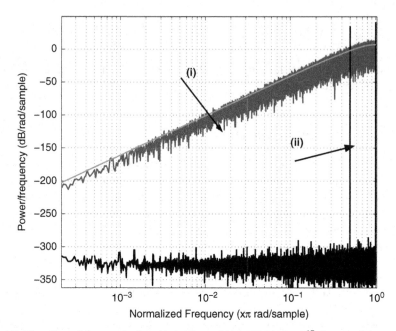

Fig. 5.3 The effect of short cycles in a third order multi-bit EFM ($M = 2^{17}$) for two different constant inputs; (*i*) $x_i[n] = 1$ and (*ii*) $x_i[n] = 2^{16}$. The *solid curve* shows shaped white quantization noise calculated from Eq. (3.8). The length of the FFT is $N_f = 2^{17}$

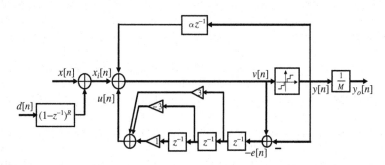

Fig. 5.4 Simulated conventional multi-level third order EFM ($\alpha = 0$ and $d[n] = 0$)

Chapter 3. Another deterministic way of increasing the cycle length is to modify the architecture [76] in such a way that the cycle length inherently attains the maximum possible value; HK-MASH belongs to this class. In [76], a first order EFM that has the maximum cycle length for *all* inputs and for *all* initial conditions has been developed. In this chapter, we show empirically that the same concept described in Chapter 4 (adding a specially chosen output feedback path αz^{-1}, as shown in Fig. 5.1) can be used to maximize the cycle lengths of higher order EFMs with a multi-bit quantizer, thereby achieving a smooth power spectrum.

Table 5.1 Corresponding values of a with respect to the modulator word length, for n_0 in the range 5–25

n_0	a
5, 7, 13, 17, 19	1
6, 9, 10, 12, 14, 20, 22, 24	3
8, 18, 25	5
11, 21	9
16, 23	15
15	19

5.2.2 HK Architecture

Consider again the EFM shown in Fig. 5.1. Let $\alpha = a/M$, where M is a power of 2 ($M = 2^{n_0}$), and $H(z) = z^{-1}$. Assume that the quantizer output y produces two levels: 0 and M. This corresponds to a conventional first-order EFM with a two-level quantizer augmented by an additional output feedback path.

The cycle length (and hence the number of tones in the spectrum) in this case is defined by

$$N = M - a$$

for *all* constant inputs and for *all* initial conditions, where a is chosen according to Table 5.1 such that $M - a$ is the largest prime number less than M.

Qualitatively, when M is a power of two and the input takes on specific (worst case) values, the quantizer output can overflow periodically, leading to very short cycles of lengths 2, 4, 8, etc. (divisors of M) [13]. The idea of the HK output feedback path is effectively to add an offset to the state so that it resets to a instead of zero when the quantizer overflows. This makes the effective quantization step equal to $M - a$ instead of M. If a is chosen to make $M - a$ prime, then *every* cycle is of length $M - a$.

In Chapter 4, it was shown that the cycle length is $(M - a)^l$ in the case of an lth order MASH structure consisting of first order EFMs with two-level quantizers. In this chapter, we consider higher order EFMs and SQ-DDSMs with multi-level quantizers. The operating principle is the same as in Chapter 4: an "overflow" at the quantizer output (in this case a level change) causes an offset to be added to the state so that the quantization interval appears to be prime.

5.3 Maximum Cycle Length EFMs

5.3.1 Architecture

In this section, we consider the HK concept applied to a higher order EFM with a multi-level quantizer. For stability reasons, higher order EFMs usually are multi-bit quantizers [5, 6]. We consider the architecture shown in Fig. 5.1 where $\alpha = a/M$

and $H(z) = 1 - (1 - z^{-1})^l$ [6]. With this filter, we obtain the STF and NTF of
Eq. (5.1). By contrast with a conventional modulator ($\alpha = 0$), a pole at $z = \alpha$
is added to both the STF and the NTF. If α is sufficiently small, this pole is very
close to the origin in the z plane; equivalently, it is a distant pole which does not
significantly affect the overall operation of the modulator.

In order to illustrate this point, consider the magnitude response of the multiplying factor $G_{hk}(z) = \frac{1}{1-\alpha z^{-1}}$ which appears in the STF and the NTF [see Eq. (5.1)].
Substituting $z = e^{j\omega}$ into $G_{hk}(z)$ and calculating the resulting magnitude (in dB)
yields

$$20\log_{10}\left(\left|G_{hk}(e^{j\omega})\right|\right) = 20\log_{10}\left(\frac{1}{\sqrt{1+\alpha^2 - 2\alpha\cos(\omega)}}\right). \qquad (5.2)$$

Figure 5.5 shows (5.2) for different values of α, namely $0, \frac{1}{2^{13}}, \frac{1}{2^7}, \frac{1}{2^5}$. As α
decreases, the resulting spectrum approaches that of $\alpha = 0$. When $\alpha = \frac{1}{2^7}$, the
magnitude of G_{hk} at low frequencies is only 0.07 dB.

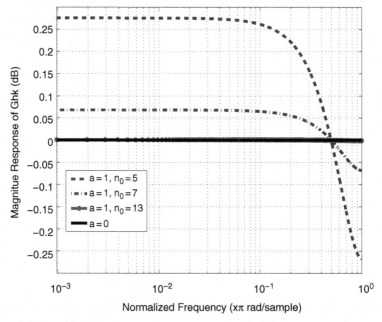

Fig. 5.5 Magnitude response of the multiplying factor $\frac{1}{1-\alpha z^{-1}}$ for four values of $\alpha = 0, \frac{1}{2^{13}}, \frac{1}{2^7}, \frac{1}{2^5}$

5.3.2 Cycle Lengths

To determine the cycles lengths, exhaustive brute force MATLAB simulations have been performed for $n_0 = 3, 4, 5$ for *all* combinations of constant inputs[1] and initial conditions, and for modulator orders $l = 1, 2, 3$. Note that we allow sufficiently many quantizer levels such that the quantizer does not overload. For example, when $l = 3$, 8 levels are allowed and when $l = 4$, 16 levels are allowed for the output swing. The constant input is swept from 0 to $M - a - 1$.

These simulations confirm that the cycle length is $(M - a)^l$ in all cases. Simulating the modulators for all possible combinations of inputs and initial conditions for orders higher than two with $n_0 \geq 6$ is almost impractical as the cycles become very long. In this case, all initial conditions were set to zero and the modulators with $n_0 = 6, 7$ were simulated for all constant inputs. Moreover, sample simulations were performed for $l = 4, 5$; all were consistent with the results in Table 5.2.

Table 5.2 compares the results of this work with the state of the art results in [11]. Note that increasing the order of the HK-EFM by one increases the cycle length by a factor of approximately 2^{n_0} compared to a factor of no more than 4 in the conventional case.

Although this result has not yet been proven mathematically, exhaustive simulations suggest that the cycle length is $(M - a)^l$ for an lth order no-overload EFM for *all* constant inputs and for *all* initial conditions, where a is determined from Table 5.1. This result is consistent with the theoretical result for the lth order HK-MASH structure with a constant input incorporating first order EFMs, each having a one-bit quantizer [76]. At the time of writing, the proof of this result is an open problem.

5.3.3 Spectral Investigation

5.3.3.1 Ditherless

In order to illustrate the effect of the maximized cycle length, Fig. 5.6 shows simulation results for a third order modulator ($l = 3$) with a step-size M equal to 2^9

Table 5.2 A comparison of minimum cycle lengths of our EFM structure and the conventional EFM [11]

Modulator order l	[11] conventional structure presetting initial conditions	HK-EFM *all* initial conditions *all* constant digital inputs
1	2^{n_0}	$(2^{n_0} - a)$
2	$2^{(n_0-1)}$	$(2^{n_0} - a)^2 \approx (2^{2n_0})$
3	$2^{(n_0+1)}$	$(2^{n_0} - a)^3 \approx (2^{3n_0})$
4	$2^{(n_0+1)}$	$(2^{n_0} - a)^4 \approx (2^{4n_0})$
5	$2^{(n_0+2)}$	$(2^{n_0} - a)^5 \approx (2^{5n_0})$

---

[1] $0, 1, 2, \cdots, M - a - 1$.

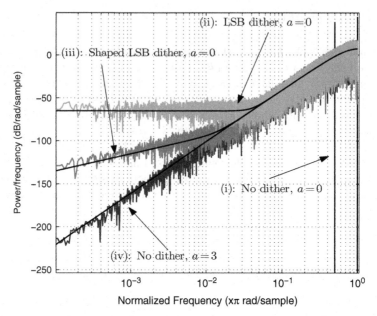

Fig. 5.6 Spectra of a third order multi-bit EFM with input decimal 256, $M = 2^9$. The *solid curves* represent the shaped white quantization noise including the dither contribution in the case of the dithered EFM. The FFT length $N_f = 2^{18}$. The DC tone has been removed

($n_0 = 9$), a constant input 256, and zero initial conditions. In this case, the modulator with non-zero normalized input 0.5 ($x[n] = 256$) yields a cycle of length 4. The effect of its short length is evident in the spectrum [see "(i): No dither, $a = 0$"], which consists of just two high power tones in the frequency range 0 to π (equivalently, 0 to $\frac{f_s}{2}$). The modulator fails to perform proper quantization noise shaping because the cycle length (4) is too small.

5.3.3.2 With Dither

A pseudo-random binary sequence is generated[2] and is added to the LSB of the modulator input. As shown in Fig. 5.6, the dither randomizes the sequences effectively. However, it degrades the low frequency part of the noise shaping spectrum by bringing up the noise floor. This is shown in the figure [see (ii): "LSB dither, $a = 0$"].

5.3.3.3 With Noise-Shaped Dither

In order to combat this effect, one can pass the dither sequence through a high pass filter such as $(1 - z^{-1})^R$ and then add the resulting signal to the input of the

---

[2] Using rounding applied to the rand function in MATLAB.

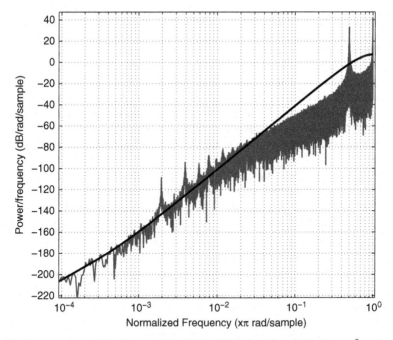

Fig. 5.7 Spectrum of a third order multi-bit EFM with decimal input 256, $M = 2^9$ and second order shaped ($R = 2$) LSB dither at the input. In this case, the shaped dither fails to perform proper randomization, resulting in a tonal spectrum at the output. The *solid curve* represents the shaped white quantization noise including the dither contribution

modulator. First order shaped LSB dither ($R = 1$) can improve the low frequency part of the spectrum because it randomizes the error sequences effectively [see "(iii): Shaped LSB dither and $a = 0$"]. Higher order shaped dither does not guarantee a spike-free spectrum, as illustrated in Fig. 5.7. The solid curve in Fig. 5.7 was plotted using the following equation:

$$S(\omega_k) = \frac{1}{M^2} \frac{1}{12} \left(2 \sin \left(\frac{\omega_k}{2} \right) \right)^2 + \frac{1}{12} \left(2 \sin \left(\frac{\omega_k}{2} \right) \right)^2, \ \omega_k \in \left\{ 0, \frac{2\pi}{N_f}, \frac{4\pi}{N_f}, \cdots, \pi \right\}.$$
(5.3)

5.3.3.4 Ditherless HK-EFM

By turning off the dither and applying the feedback path αz^{-1} to the system with $a = 3$ chosen from Table 5.1, a smooth spike-free noise-shaping spectrum with a slope of 60 dB/decade is achieved [see "(iv): No dither, $a = 3$"], corresponding to the idealized prediction obtained in the literature by assuming additive white noise for the quantizer error [4]. In this case, the low frequency part of the spectrum is not degraded and the HK-EFM outperforms the shaped LSB dithering technique.

5.4 Maximum Cycle Length Single-Quantizer DDSMs

5.4.1 Architecture

The idea described in Section 5.3 can be applied to another class of digital modulators with a similar STF and NTF. The generic block diagram of the single-quantizer DDSM (SQ-DDSM) with the auxiliary output feedback path α and with noise-shaped LSB dither is shown in Fig. 5.8. In this case, there is already a delay in the forward path so it is not necessary to introduce additional delay in the α feedback path.

In the forward path, the block $F(z)$ filters the signal u and delivers v to the input of the multi-bit quantizer. The output of the multi-bit quantizer y, which is an integer multiple of M, is fed back to the input summing node via filter block $G(z)$.

As in the case of the maximum cycle length HK-EFM, the block α has been added to the modulator, as a deterministic alternative to dithering, in order to randomize the quantizer error sequence e.

For this modulator, $STF(z)$ and $NTF(z)$ can be written in terms of $F(z)$, $G(z)$ and α as follows:

$$STF(z) = \frac{F(z)}{1 + F(z)G(z) - \alpha F(z)},$$

$$NTF(z) = \frac{1}{1 + F(z)G(z) - \alpha F(z)}. \tag{5.4}$$

$F(z)$ and $G(z)$ are chosen such that STF has a low pass or all pass characteristic (passing the low frequency input signal) and NTF has a high pass characteristic, rejecting the quantization noise at low frequencies. We consider a case [61] where $F(z)$ and $G(z)$ are of the form

$$F(z) = z^{-l}(1 - z^{-1})^{-l}; \quad G(z) = z^l - (z-1)^l, \tag{5.5}$$

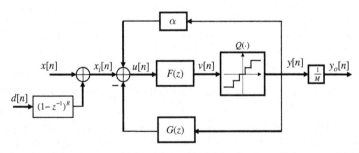

Fig. 5.8 Generic block diagram of SQ-DDSM with the feedback coefficient α and with shaped LSB dither

giving

$$STF(z) = \frac{z^{-l}}{1 - \alpha z^{-l}}; \quad NTF(z) = \frac{(1 - z^{-1})^l}{1 - \alpha z^{-l}}. \tag{5.6}$$

In the following, we demonstrate the improvement in the spectral performance of the modulator resulting from the feedback block α.

5.4.2 Simulation Results

Similar to the cases of the HK-EFM and the HK-MASH [76], extensive simulations suggest that the cycle length is $(M - a)^l$ for an lth order no-overload modulator for all constant inputs and for all initial conditions, although this has not been proven.

Figure 5.9 shows the simulation results for the third order modulator ($l = 3$) in Fig. 5.10 with the step-size M equal to 2^9, a constant input 256, $N_{\min} = 4$, $N_{\max} = 4$ and zero initial conditions on the internal registers.

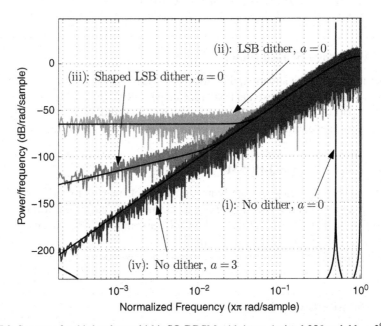

Fig. 5.9 Spectra of a third order multi-bit SQ-DDSM with input decimal 256 and $M = 2^9$. The *solid curves* represent the shaped white quantization noise including the dither contribution in the case of the dithered SQ-DDSM with higher order noise shaping. The length of the FFT is $N_f = 2^{18}$. The DC tone has been removed

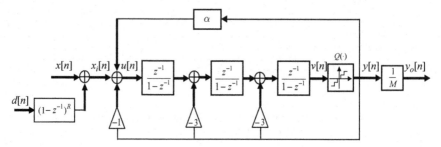

Fig. 5.10 Third order SQ-DDSM with additional feedback block α and with noise-shaped dither at the input

5.4.2.1 Ditherless

In this case, the modulator with non-zero normalized input 0.5 ($x[n] = 256$) yields a cycle of length 4. This effect is clear in the spectrum [see: (i): "No dither, $a = 0$"]. Once again, there are only two high power tones in the frequency range 0 to π.

5.4.2.2 With White Dither

A one-bit pseudo-random dither signal is added to the LSB of the modulator input [61]. The low frequency part of the spectrum is degraded due to the noise floor introduced by the dither [see (ii): "LSB dither, $a = 0$"].

5.4.2.3 With Noise-Shaped Dither

In order to improve the situation, the dither is passed through a filter $(1 - z^{-1})^R$ in this case. Simulations suggest that first order shaped LSB dither can improve the low frequency part of the spectrum [see (iii): "Shaped LSB dither, $a = 0$"]. However, higher order shaped dither again does not guarantee a spike-free spectrum.

5.4.2.4 Ditherless HK-SQ-DDSM

By turning off the dither and applying the feedback path α to the system with $a = 3$, a smooth spike-free noise-shaping spectrum with a slope of 60 dB/decade is achieved, as predicted by the additive white noise assumption for the quantizer error. In this case, the low frequency part of the spectrum is not degraded and it outperforms the shaped LSB dithering technique [see (iv): "No dither, $a = 3$"].

5.5 Implementation Issues

When the modulator is implemented digitally, the step-size of the quantizer is usually chosen, for simplicity, to be a power of two. The coefficient α should be chosen such that $M - a$ is the largest prime number less than M, where M is the step-size of the quantizer and a is a small integer reported in Table 5.1.[3] With this selection, only a small fraction of the output signal y is fed back to the input. The effect of the coefficient α on the noise and signal transfer functions is negligible for both HK-EFMs and HK-SQ-DDSMs when $M > 2^4$.

In terms of hardware complexity, the output y can be scaled by $1/M$ to reduce the number of output bits significantly; this corresponds to discarding zeroed LSBs when M is a power or 2. The additional feedback path requires a multiplication by multiples of small integers from Table 5.1; this could be implemented by a multiplexer. An additional adder is required at the input summing node but the number of bits is small; it is similar in complexity to that required for noise-shaped LSB dither. Overall, the complexity of each modified structure is marginally less than the corresponding dithered conventional solution because neither the PRBS generator nor the filter is required.

The stable dynamic range of a higher order multi-bit modulator is determined primarily by the quantizer. Since the effect of the additional feedback path is equivalent to a minor perturbation of the quantization step, the effect on the stable dynamic range is also negligible. In other words, the stable dynamic range is approximately equivalent to that of the conventional ($a = 0$) modulator.

5.6 Summary

This chapter shows how the HK feedback idea described in Chapter 4 can be applied to two other classes of DDSMs: EFM and SQ-DDSM. Extensive simulations suggest that adding the feedback path α introduced in Chapter 4 results in maximized cycle length DDSMs. Also, the exact values of cycle lengths are determined by the same formula as the HK-MASH, namely $(M - a)^l$, where a is the small integer chosen from Table 5.1 and l is the order of the modulator.

The spectral performance of the modified modulators is compared with those of conventional dithered modulators. Simulations show that HK-EFM and HK-SQ-DDSM yield better low frequency performance compared to their conventional dithered counterparts.

Unlike in the case of HK-MASH, the formula $(M - a)^l$ for cycle lengths for the modulators studied in this chapter has not been proven mathematically at the time of writing.

---

[3] The range of n_0 given in the table can be extended beyond 25 by choosing a such that $M - a$ is the largest prime number less than M.

Appendix A
Calculating the Mean and Variance of the Error Signal in Mid-Tread and 1-Bit Quantizers

A.1 Mid-Tread Quantizer

In the digital mid-tread quantizer considered in Section 3.2.1 with the characteristic shown in Fig. A.1, the error signal in the no overload range is in the range $[-\frac{M}{2} + 1, \frac{M}{2}]$, as shown in Fig. A.2. If the error signal is uniformly distributed within this range, as shown in Fig. A.3, we can calculate its mean and variance as follows.

Fig. A.1 The input–output characteristic of a multi-level digital mid-tread quantizer with 5 output levels

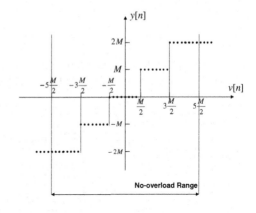

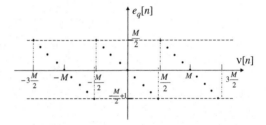

Fig. A.2 Illustration of the error signal e_q for a mid-tread quantizer like that shown in Fig. A.1

K. Hosseini, M.P. Kennedy, *Minimizing Spurious Tones in Digital Delta-Sigma Modulators*, Analog Circuits and Signal Processing, DOI 10.1007/978-1-4614-0094-3, © Springer Science+Business Media, LLC 2011

Fig. A.3 Probability mass
function of a uniformly
distributed random variable
in the range $[-\frac{M}{2}+1, \frac{M}{2}]$

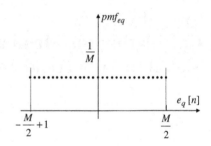

The mean is defined by

$$
\begin{aligned}
M_e = E[e_q] &= \sum_{k=\frac{-M}{2}+1}^{\frac{M}{2}} kP(e_q = k) \\
&= \frac{1}{M} \sum_{k=\frac{-M}{2}+1}^{\frac{M}{2}} k \\
&= \frac{1}{2},
\end{aligned}
\tag{A.1}
$$

where $P(e_q = k)$ denotes the probability that $e_q = k$.

In order to find the variance, we first calculate $E[e_q^2]$:

$$
\begin{aligned}
E[e_q^2] &= \sum_{k=\frac{-M}{2}+1}^{\frac{M}{2}} k^2 P(e_q = k) \\
&= \frac{1}{M} \sum_{k=\frac{-M}{2}+1}^{\frac{M}{2}} k^2 \\
&= \frac{1}{M}\left(\frac{M}{2}\right)^2 + 2\frac{1}{M}\sum_{k=0}^{\frac{M}{2}-1} k^2 \\
&= \frac{M}{4} + 2\frac{1}{M}\frac{\left(\frac{M}{2}-1\right)\left(\frac{M}{2}\right)(M-1)}{6} \\
&= \frac{M^2+2}{12}.
\end{aligned}
\tag{A.2}
$$

In deriving (A.2), we have used the identity:

$$\sum_{k=0}^{M-1} k^2 = \frac{(M-1)(M)(2M-1)}{6}. \tag{A.3}$$

We can use Eqs. (A.1) and (A.2) to find the variance, which is given by

$$\sigma_{ee}^2 = E[e_q^2] - (E[e_q])^2 = \frac{M^2 - 1}{12}. \tag{A.4}$$

A.2 1-Bit Quantizer

We consider the quantizer shown in Fig. 2.10c. The quantization error signal e_q is in the range $\left\{0, -\frac{1}{M}, \cdots, -\frac{M-1}{M}\right\}$. Assuming a uniform distribution for e_q one can obtain:

$$E[e_q] = -\frac{1}{2}\left(\frac{M-1}{M}\right). \tag{A.5}$$

For large M, (A.5) is approximately -0.5.

In order to determine the variance, we calculate $E[e_q^2]$:

$$E[e_q^2] = \frac{1}{M^2} \sum_{k=0}^{M-1} k^2 P(e_q = k)$$

$$= \frac{1}{M^3} \sum_{k=0}^{M-1} k^2$$

$$= \frac{(M-1)(2M-1)}{6M^2}. \tag{A.6}$$

Using (A.6) and (A.5), we obtain the variance:

$$\sigma_{e_q}^2 = E[e_q^2] - (E[e_q])^2 = \frac{M^2 - 1}{12M^2}. \tag{A.7}$$

For large M, (A.7) is approximately $\frac{1}{12}$.

Appendix B
Mathematical Analysis of the HK-MASH DDSM

B.1 Proof of the Cycle Length for the Modified First Order Delta Sigma Modulator

Consider the modified first order EFM in Fig. 4.1a which has the maximum cycle length for *all* constant digital inputs and for *all* initial conditions. For this structure, we have that

$$e[n] = v[n] \bmod M, \tag{B.1}$$

where $M = 2^{n_0}$ and n_0 is the accumulator word length. Considering Fig. 4.1a, note that we can write (B.1) as:

$$e[n] = \left(x[n] + e[n-1] + ay[n-1]\right) \bmod M. \tag{B.2}$$

Expanding (B.2) with its indices, we have:

$$e[0] = \left(x[0] + s[0]\right) \bmod M, \tag{B.3}$$

where $s[0]$ is the initial condition. Continuing with the other indices, we obtain:

$$e[1] = \left(x[1] + e[0] + ay[0]\right) \bmod M, \tag{B.4}$$
$$e[2] = \left(x[2] + e[1] + ay[1]\right) \bmod M, \tag{B.5}$$

and for the indices $(n-1)$ and n we have:

$$e[n-1] = \left(x[n-1] + e[n-2] + ay[n-2]\right) \bmod M, \tag{B.6}$$
$$e[n] = \left(x[n] + e[n-1] + ay[n-1]\right) \bmod M. \tag{B.7}$$

If we replace the value of each $e[i]$ with its value from the previous equation and recall that:

$$\left(a + b \bmod M\right)\bmod M = \left(a + b\right)\bmod M, \tag{B.8}$$

for positive integers a and b [5], we can write $e[n]$ as:

$$e[n] = \left(s[0] + \sum_{k=0}^{n} x[k] + a \sum_{k=0}^{n-1} y[k] \right) \bmod M, \tag{B.9}$$

which, in the case of a constant dc input X, gives:

$$e[n] = \left(s[0] + (n+1)X + a \sum_{k=0}^{n-1} y[k] \right) \bmod M. \tag{B.10}$$

Before starting the procedure of determining the period of (B.9), we find a relationship between $x[n]$ and $y[n]$. By looking at Fig. 4.1a, we write:

$$\begin{aligned} e[n] &= v[n] - My[n], \\ &= ay[n-1] + x[n] + e[n-1] - My[n], \end{aligned} \tag{B.11}$$

which we can rewrite as:

$$My[n] - ay[n-1] = x[n] + e[n-1] - e[n]. \tag{B.12}$$

Expanding (B.12) with its indices yields:

$$My[1] - ay[0] = x[1] + e[0] - e[1], \tag{B.13}$$
$$My[2] - ay[1] = x[2] + e[1] - e[2], \tag{B.14}$$

$$\vdots$$

$$My[n] - ay[n-1] = x[n] + e[n-1] - e[n]. \tag{B.15}$$

If we add the terms on the left side together and do the same for the terms on the right side of the above equations, we obtain:

$$\sum_{k=1}^{n} My[k] - a \sum_{k=0}^{n-1} y[k] = \sum_{k=1}^{n} x[k] + e[0] - e[n]. \tag{B.16}$$

If the system is periodic in the steady state with period N, we have:

$$\sum_{k=1}^{N} My[k] - a \sum_{k=0}^{N-1} y[k] = \sum_{k=1}^{N} x[k] + e[0] - e[N], \tag{B.17}$$

where $e[0] = e[N]$, and

$$\sum_{k=1}^{N} y[k] = \sum_{k=0}^{N-1} y[k].$$ (B.18)

Therefore, in the steady state, we obtain:

$$(M - a) \sum_{k=1}^{N} y[k] = \sum_{k=1}^{N} x[k].$$ (B.19)

Rearranging (B.19) gives

$$\sum_{k=1}^{N} y[k] = \frac{1}{M - a} \sum_{k=1}^{N} x[k].$$ (B.20)

For a constant DC input X, we have:

$$\sum_{k=1}^{N} y[k] = \frac{NX}{M - a}.$$ (B.21)

Now we return to find N, the period of the system. In order for (B.10) to be periodic with period N, we require that:

$$e[n] = e[n + N].$$ (B.22)

Substituting from (B.10) for $e[n]$ and $e[n + N]$ in (B.22), we obtain:

$$\left(NX + a \sum_{k=1}^{N} y[k]\right) \mod M = 0,$$ (B.23)

which is independent of the initial condition, $s[0]$. Using (B.21) in (B.23), we obtain:

$$\left(NX + a\frac{NX}{M - a}\right) \mod M = 0,$$

$$\left(\frac{NX}{M - a}\right) M \mod M = 0.$$ (B.24)

If we choose $M - a$ to be prime and if $0 < X < (M - a)$, then the minimum non-zero solution for N such that the above equation is valid is:

$$N = M - a, \text{ where } N \text{ is prime.}$$ (B.25)

This solution is correct, regardless of the input X and the initial condition.

The difference between M and the closest prime number to it is a, the values of which, for different modulator word lengths n_0 ($5 \leq n_0 \leq 25$), are shown in Table 4.1.

B.2 Proof of the Cycle Length for the HK-MASH Modulator

We extend the results of the previous section to find the cycle length of the higher order HK-MASH modulator shown in Fig. 4.3. We have proven that, for the first order modulator, the cycle length for all digital inputs and for all initial conditions is N [see (B.25)], where N is prime. This imposes a restriction on the form of the period of the higher order modulators. If the higher order modulator is periodic with a period N_l, then we require that

$$N_l = KN, \tag{B.26}$$

where N is prime and is determined by (B.25), and K is an integer.

B.2.1 Modified Second Order Modulator

We use Eq. (B.20) to find the cycle length N_2 of a second order modulator. Considering Fig. 4.3, and using (B.25) and (B.26), we write:

$$\sum_{k=1}^{N_2} y_2[k] = \left(\frac{1}{M-a}\right) \sum_{k=1}^{N_2} e_1[k],$$

$$\sum_{k=1}^{KN} y_2[k] = \left(\frac{K}{M-a}\right) \sum_{k=1}^{N} e_1[k],$$

$$= \left(\frac{K}{M-a}\right) \sum_{k=1}^{N} e_1[k],$$

$$= \left(\frac{K}{N}\right) \sum_{k=1}^{N} e_1[k]. \tag{B.27}$$

The only solution for K so that the right side is an integer is N, provided that the greatest common divisor of N and $\sum_{k=1}^{N} e_1[k]$ is 1. This condition is true; otherwise the solution for K is not unique. With this solution for K, (B.26) yields:

$$N_2 = N^2, \tag{B.28}$$

where N is prime and is determined by (B.25).

B.2.2 Modified Third Order and Higher Order Modulators

The solution found in the previous section is used in this section to find the cycle length of the third stage. The solution for N_3 should be in the form of:

$$N_2 = KN^2, \tag{B.29}$$

where N is prime and K is an integer. Rewriting (B.20) for the third stage and repeating the same procedure performed for the second stage, we have:

$$\sum_{k=1}^{KN^2} y_3[k] = \frac{K}{N} \sum_{k=1}^{N^2} e_2[k]. \tag{B.30}$$

With the same reasoning as for the second stage, the solution for K is N and therefore we have:

$$N_3 = N^3, \tag{B.31}$$

where N is prime and determined by (B.25).

With the same reasoning as for the second and third stage cases, the solution for the lth order modulator is:

$$N_l = N^l, \tag{B.32}$$

where N is prime and determined by (B.25). Adding an extra stage to this modified structure increases the cycle length by a factor of N, independently of the initial condition and the constant input.

References

1. F. De Jager, "Delta modulation-a method of PCM transmission using the one unit code," *Philips Research Report*, vol. 7, pp. 442–466, 1952.
2. C. C. Cutler, "Transmission system employing quantization," U.S. Patent 2,927,962, 1960.
3. H. Inose, Y. Yasuda, and J. Murakami, "A telemetering system by code manipulation-Δ-Σ modulation," *IRE Transactions on Space Electronics and Telemetry*, vol. SET-8, pp. 204–209, Sept. 1962.
4. S. R. Norsworthy, R. Schreier, and G. C. Temes, *Delta-Sigma Data Converters: Theory, Design, and Simulaion*. New York: IEEE Press, 1997.
5. M. Kozak and I. Kale, *Oversampled Delta-Sigma Modulators, Analysis, Applications and Novel Topologies*. Boston: Kluwer Academic Publishers, 2003.
6. R. Schreier and G. C. Temes, *Understanding Delta-Sigma Data Converters*. New Jersey: IEEE Press, 2004.
7. A. V. Oppenheim, R. W. Schafer, and J. R. Buck, *Discrete-Time Signal Processing*. Upper Saddle River, NJ: Prentice-Hall, 1999.
8. D. Reefman, J. Reiss, E. Janssen, and M. Sandler, "Description of limit cycles in sigma-delta modulators," *IEEE Transactions on Circuits and Systems-I: Regular Papers*, vol. 52, no. 6, pp. 1211–1223, June 2005.
9. S. Pamarti and I. Galton, "LSB dithering in MASH delta-sigma D/A converters," *IEEE Transactions on Circuits and Systems I*, vol. 54, no. 4, pp. 779–790, Apr. 2007.
10. J. Reiss and M. Sandler, "Detection and removal of limit cycles in sigma delta modulators," *IEEE Transactions on Circuits and Systems I: Regular Papers*, vol. 55, no. 10, pp. 3119–3130, 2008.
11. M. J. Borkowski, T. A. D. Riley, J. Hakkinen, and J. Kostamovaara, "A practical delta sigma modulator design method based on periodical behavior analysis," *IEEE Transactions on Circuits and Systems II: Express Briefs*, vol. 52, pp. 626–630, Oct. 2005.
12. M. Borkowski, *Digital $\Delta\Sigma$ Modulation: Variable Modulus and Tonal Behavior in a Fixed-Point Digital Environment*. Oulu: Oulu University Press, 2008.
13. K. Hosseini and M. P. Kennedy, "Mathematical analysis of digital MASH delta-sigma modulators for fractional-N frequency synthesizers," in *Proceedings of PRIME 2006*, Lecce, Italy, June 2006, pp. 309–312.
14. K. Hosseini and M. P. Kennedy, "Calculation of sequence lengths in MASH 1-1-1 digital delta sigma modulators with a constant input," in *Proceedings of PRIME 2007*, Bordeaux, France, July 2007, pp. 13–16.
15. P. Level and S. R. L. Camino, "Digital to digital sigma-delta modulator and digital frequency synthesizer incorporating the same," US Patent 6,822,593 B2, Nov. 23, 2004.
16. K. Hosseini and M. P. Kennedy, "Mathematical analysis of a prime modulus quantizer MASH digital delta-sigma modulator," *IEEE Transactions on Circuits and Systems Part II: Express Briefs*, vol. 54, no. 12, pp. 1105–1109, Dec. 2007.
17. R. Gray and D. Neuhoff, "Quantization," *IEEE Transactions on Information Theory*, vol. 44, no. 6, pp. 2325–2383, 1998.

18. G. Mitteregger et al., "A 14b 20mW 640MHz CMOS CT ADC with 20MHz signal band-width and 12b ENOB," in *Digest of Technical Papers, IEEE International Solid-State Circuits Conference*, 2006, pp. 131–132.

19. L. Breems et al., "A 56mW CT quadrature cascaded $\Delta\Sigma$ modulator with 77dB DR in a near zero-IF 20MHz band," in *Digest of Technical Papers, IEEE International Solid-State Circuits Conference*, Feb. 2007, pp. 238–599.

20. W. Yang, W. Schofield, H. Shibata, S. Korrapati, A. Shaikh, N. Abaskharoun, and D. Ribner, "A 100mW 10MHz-BW CT $\Delta\Sigma$ modulator with 87dB DR and 91dBc IMD," in *Digest of Technical Papers, IEEE International Solid-State Circuits Conference*, 2008, pp. 498–631.

21. Y. Shu, B. Song, and K. Bacrania, "A 65nm CMOS CT $\Delta\Sigma$ modulator with 81dB DR and 8MHz BW auto-tuned by pulse injection," in *Digest of Technical Papers, IEEE International Solid-State Circuits Conference,* 2008, pp. 500–501.

22. S. Ouzounov, E. Roza, J. Hegt, G. van der Weide, and A. van Roermund, "Analysis and design of high-performance asynchronous sigma-delta modulators with a binary quantizer," *IEEE Journal of Solid-State Circuits*, vol. 41, no. 3, pp. 588–596, Mar. 2006.

23. E. Roza, "Analog-to-digital conversion via duty-cycle modulation," *IEEE Transactions on Circuits and Systems II: Analog and Digital Signal Processing*, vol. 44, no. 11, pp. 907–914, 1997.

24. J. Daniels, W. Dehaene, M. Steyaert, and A. Wiesbauer, "A/D conversion using an asynchronous delta-sigma modulator and a time-to-digital converter," in *Proceedings of ISCAS 2008, IEEE International Symposium on Circuits and Systems*, May 2008, pp. 1648–1651.

25. P. Malla, H. Lakdawala, K. Kornegay, and K. Soumyanath, "A 28mW spectrum-sensing re-configurable 20MHz 72dB-SNR 70dB-SNDR DT $\Delta\Sigma$ ADC for 802.11 n/WiMAX receivers," in *Digest of Technical Papers, IEEE International Solid-State Circuits Conference*, 2008, pp. 496–631.

26. Y. Chae, I. Lee, and G. Han, "A 0.7 V 36 μW 85 dB-DR audio $\Delta\Sigma$ modulator using class-C inverter," in *Digest of Technical Papers, IEEE International Solid-State Circuits Conference*, 2008, pp. 490–630.

27. T. Hamasaki, Y. Shinohara, H. Terasawa, K. Ochiai, M. Hiraoka, and H. Kanayama, "A 3-V, 22-mW multibit current-mode DAC with 100dB dynamic range," *IEEE Journal of Solid-State Circuits*, vol. 31, no. 12, pp. 1888–1894, 1996.

28. R. Adams and K. Nguyen, "A 113-dB SNR oversampling DAC with segmented noise-shaped scrambling," *IEEE Journal of Solid-State Circuits*, vol. 33, no. 12, pp. 1871–1878, 1998.

29. I. Fujimori and T. Sugimoto, "A 1.5 V, 4.1 mW dual-channel audio delta-sigma D/A converter," *IEEE Journal of Solid-State Circuits*, vol. 33, no. 12, pp. 1863–1870, 1998.

30. I. Fujimori, A. Nogi, and T. Sugimoto, "A multibit delta-sigma audio DAC with 120-dB dynamic range," *IEEE Journal of Solid-State Circuits*, vol. 35, no. 8, pp. 1066–1073, 2000.

31. M. Annovazzi, V. Colonna, G. Gandolfi, F. Stefani, and A. Baschirotto, "A low-power 98-dB multibit audio DAC in a standard 3.3-V 0.35-μm CMOS technology," *IEEE Journal of Solid-State Circuits*, vol. 37, no. 7, pp. 825–834, 2002.

32. V. Colonna, M. Annovazzi, G. Boarin, G. Gandolfi, F. Stefani, and A. Baschirotto, "A 0.22-mm² 7.25-mW per-channel audio stereo-DAC with 97-dB DR and 39-dB SNRout," *IEEE Journal of Solid-State Circuits*, vol. 40, no. 7, pp. 1491–1498, 2005.

33. K. Nguyen, A. Bandyopadhyay, B. Adams, K. Sweetland, and P. Baginski, "A 108dB SNR 1.1 mW oversampling audio DAC with a three-level DEM technique," in *Digest of Technical Papers, IEEE International Solid-State Circuits Conference*, 2008, pp. 488–630.

34. M. Perrott et al., "A 27-mW CMOS fractional-N synthesizer using digital compensation for 2.5-Mb/s GFSK modulation," *IEEE Journal of Solid-State Circuits*, vol. 32, no. 12, pp. 2048–2060, 1997.

35. W. Rhee, B. Song, and A. Ali, "A 1.1-GHz CMOS fractional-N frequency synthesizer with a 3-b third-order $\Delta\Sigma$ modulator," *IEEE Journal of Solid-State Circuits*, vol. 35, no. 10, pp. 1453–1460, 2000.

36. S. Wilingham, M. Perrott, B. Setterberg, A. Grzegorek, and B. McFarland, "An integrated 2.5 GHz $\Sigma\Delta$ frequency synthesizer with 5 μs settling and 2 Mb/s closed loop modulation,"

in *Digest of Technical Papers, IEEE International Solid-State Circuits Conference,* 2000, pp. 200–201.

37. B. D. Muer and M. Steyaert, "A CMOS monolithic $\Delta\Sigma$ controlled fractional-N frequency synthesizer for DCS-1800," *IEEE Journal of Solid-State Circuits,* vol. 37, no. 7, pp. 835–844, July 2002.

38. S. Pamarti, L. Jansson, and I. Galton, "A wideband 2.4-GHz $\Delta\Sigma$ fractional-N PLL with 1 Mb/s in-loop modulation," *IEEE Journal of Solid-State Circuits,* vol. 39, no. 1, pp. 49–62, Jan. 2004.

39. S. Meninger and M. Perrott, "A 1-MHZ bandwidth 3.6-GHz 0.18-μm CMOS fractional-N synthesizer utilizing a hybrid PFD/DAC structure for reduced broadband phase noise," *IEEE Journal of Solid-State Circuits,* vol. 41, no. 4, pp. 966–980, 2006.

40. S. Pellerano, S. Levantino, C. Samori, and A. Lacaita, "A dual-band frequency synthesizer for 802.11 a/b/g with fractional-spur averaging technique," in *Digest of Technical Papers, IEEE International Solid-State Circuits Conference,* 2005, pp. 104–106.

41. A. Swaminathan, K. Wang, and I. Galton, "A wide-bandwidth 2.4 GHz ISM band fractional-N PLL with adaptive phase noise cancellation," *IEEE Journal of Solid-State Circuits,* vol. 42, no. 12, pp. 2639–2650, 2007.

42. E. Temporiti, G. Albasini, I. Bietti, R. Castello, and M. Colombo, "A 700-kHz bandwidth $\Sigma\Delta$ fractional synthesizer with spurs compensation and linearization techniques for WCDMA applications," *IEEE Journal of Solid-State Circuits,* vol. 39, no. 9, pp. 1446–1454, 2004.

43. T. Riley and M. Copeland, "A simplified continuous phase modulator technique," *IEEE Transactions on Circuits and Systems II: Analog and Digital Signal Processing,* vol. 41, no. 5, pp. 321–328, 1994.

44. B. D. Muer and M. Steyaert, *CMOS Fractional-N Synthesizers: Design for High Spectral Purity and Monolithic Integration.* Boston: Kluwer Academic Publishers, 2003.

45. B. Razavi, Ed., *Phase-Locking in High-Performance Systems.* New York: IEEE Press, 2003.

46. K. Lee et al., "A 0.8 V, 2.6 mW, 88 dB dual-channel audio delta-sigma D/A converter with headphone driver," *IEEE Journal of Solid-State Circuits,* vol. 44, no. 3, pp. 916–927, Mar. 2009.

47. B. Razavi, "Challenges in the design of frequency synthesizers for wireless applications," in *Proceedings of IEEE CICC 1997, IEEE Custom Integrated Circuits Conference,* May 1997, pp. 395–402.

48. J. Rogers, C. Plett, and F. Dai, *Integrated Circuit Design for High-speed Frequency Synthesis.* Boston: Artech House, 2006.

49. B. Razavi, *Design of Analog CMOS Integrated Circuits.* Boston: McGraw-Hill, 2001.

50. F. Gardner, "Charge-pump phase-locked loops," *IEEE Transactions on Communications,* vol. 28, no. 11, pp. 1849–1858, 1980.

51. Z. Ye, "Modelling, simulation and architecture modification of delta-sigma fractional-N frequency synthesizers," Ph.D. dissertation, University College Cork, 2008.

52. C. A. Kingsford-Smith, *Patent No. 3,928,813.* Washington, DC: US Patent Office, 1975.

53. U. Rohde and J. Wiley, *Microwave and Wireless Synthesizers: Theory and Design.* Colorado: Wiley, 1997.

54. B. Miller and B. Conley, "A multiple modulator fractional divider," in *Proceedings of the 44th Annual Symposium on Frequency Control,* 1990, pp. 559–568.

55. B. Miller and R. J. Conley, "A multiple modulator fractional divider," *IEEE Transactions on Instrumentation and Measurement,* vol. 40, no. 3, pp. 578–583, 1991.

56. T. A. D. Riley, M. A. Copland, and T. A. Kwasniewski, "Delta-sigma modulation in fractional-N frequency synthesis," *IEEE Journal of Solid-State Circuits,* vol. 28, no. 5, pp. 553–559, May 1993.

57. Z. Ye and M. P. Kennedy, "Modeling and simulation of fractional-N PLL frequency synthesizer in Verilog-AMS," *Transactions on IEICE,* vol. E90-A, no. 10, pp. 2141–2147, Oct. 2007.

58. M. J. Borkowski and J. Kostamovaara, "On randomization of digital delta-sigma modulators with DC inputs," in *Proceedings of ISCAS 2006,* Kos, Greece, May 2006, pp. 3770–3773.

59. J. C. Candy and G. C. Temes, *Oversampling Delta Sigma Data Converters Theory, Design and Simulation.* New York: IEEE Press, 1992.

60. I. Galton, "One-bit dithering in delta-sigma modulator-based D/A conversion," in *ISCAS'93, IEEE International Symposium on Circuits and Systems*, vol. 2, pp. 1310–1313, 1993.
61. S. Pamarti, J. Welz, and I. Galton, "Statistics of the quantization noise in 1-bit dithered single-quantizer digital delta-sigma modulators," *IEEE Transactions on Circuits and Systems-I: Regular Papers*, vol. 54, no. 3, pp. 492–503, Mar. 2007.
62. M. Annovazzi et al., "A low-power 98-dB multibit audio DAC in a standard 3.3-V 0.35 μm CMOS technology," *IEEE Journal of Solid-State Circuits*, vol. 37, no. 7, pp. 825–834, July 2002.
63. S. Willingham et al., "An integrated 2.5-GHz $\Delta\Sigma$ frequency synthesizer with 5 μs settling and 2 Mb/s closed loop modulation," in *Digest of Technical Papers, IEEE International Solid-State Circuits Conference*, vol. 43, pp. 200–201, 2000.
64. H. Hsieh and C.-L. Lin, "Spectral shaping of dithered quantization errors in sigma-delta modulators," *IEEE Transactions on Circuits and Systems-I: Regular Papers*, vol. 54, no. 5, pp. 972–980, May 2007.
65. S. R. Norsworthy, "Effective dithering of delta-sigma modulators," in *Proceedings of ISCAS'92*, vol. 3, pp. 1304–1307, May 1992.
66. W. Chou, "Sigma delta and multi-stage sigma delta modulation with inside loop dithering," in *Proceedings of International Conference on Acoustics, Speech, and Signal Processing*, vol. 3, pp. 1953–1956, Apr. 1991.
67. R. M. Gray, "Spectral analysis of quantization noise in a single-loop sigma-delta modulator with DC input," *IEEE Transactions on Communications*, vol. 37, pp. 588–599, Jun. 1989.
68. P. W. M. Wong and R. M. Gray, "Two-stage sigma-delta modulation," *IEEE Transactions on Acoustics, Speech, & Signal Processing*, vol. 38, pp. 1937–1952, Nov. 1990.
69. W. Chua, P. W. M. Wong, and R. M. Gray, "Multi-stage sigma-delta modulation," *IEEE Transactions of Information Theory*, vol. 35, pp. 784–796, July 1989.
70. N. He, F. Kuhlmann, and A. Buzo, "Double-loop sigma-delta modulation with DC input," *IEEE Transactions on Communications*, vol. COM-38, pp. 487–495, Apr. 1990.
71. N. He, F. Kuhlmann, and A. Buzo, "Multiloop sigma-delta quantization," *IEEE Transactions of Information Theory*, vol. 38, pp. 1015–1028, May 1992.
72. B. Fitzgibbon and M. P. Kennedy, "Calculation of cycle lengths in higher-order MASH DDSMs with constant inputs," in *Proceedings of ICECS*, Athens, Greece, May 2010, pp. 484–487.
73. B. Fitzgibbon and M. P. Kennedy, "Calculation of cycle lengths in higher-order error feedback modulators with constant inputs," *IEEE Transactions on Circuits and Systems II: Express Briefs*, vol. 58, pp. 6–10, Jan. 2011.
74. *Datasheet: Analog Devices Part ADF4193*. http://www.analog.com/static/imported-files/data_sheets/ADF4193.pdf
75. J. Song and I.-C. Park, "Spur-free MASH delta-sigma modulation," *IEEE Transactions on Circuits and Systems I: Regular Papers*, vol. 57, no. 9, pp. 2426–2437, Sept. 2010.
76. K. Hosseini and M. P. Kennedy, "Maximum sequence length MASH digital delta sigma modulators," *IEEE Transactions on Circuits and Systems I*, vol. 54, no. 12, pp. 2628–2638, Dec. 2007.
77. K. Hosseini and M. P. Kennedy, "A sigma-delta modulator," Filed: Irish Patent NATI82/P/IE, Oct. 4, 2006.
78. http://www.xilinx.com/
79. http://www.analog.com/
80. http://en.wikipedia.org/wiki/Linear_feedback_shift_register
81. K. Hosseini and M. P. Kennedy, "Architectures for maximum sequence length digital delta-sigma modulators," *IEEE Transactions on Circuits and Systems II: Express Briefs*, vol. 55, no. 10, pp. 1104–1108, Nov. 2008.

Index